The Climate C

Ross Michael Pink

The Climate Change Crisis

Solutions and Adaption for a Planet in Peril

Ross Michael Pink
Kwantlen Polytechnic University (Surrey)
Burnaby, BC, Canada

ISBN 978-3-030-10008-7   ISBN 978-3-319-71033-4 (eBook)
https://doi.org/10.1007/978-3-319-71033-4

© The Editor(s) (if applicable) and The Author(s) 2018
Softcover re-print of the Hardcover 1st edition 2018
This work is subject to copyright. All rights are solely and exclusively licensed by the Publisher, whether the whole or part of the material is concerned, specifically the rights of translation, reprinting, reuse of illustrations, recitation, broadcasting, reproduction on microfilms or in any other physical way, and transmission or information storage and retrieval, electronic adaptation, computer software, or by similar or dissimilar methodology now known or hereafter developed.
The use of general descriptive names, registered names, trademarks, service marks, etc. in this publication does not imply, even in the absence of a specific statement, that such names are exempt from the relevant protective laws and regulations and therefore free for general use.
The publisher, the authors and the editors are safe to assume that the advice and information in this book are believed to be true and accurate at the date of publication. Neither the publisher nor the authors or the editors give a warranty, express or implied, with respect to the material contained herein or for any errors or omissions that may have been made. The publisher remains neutral with regard to jurisdictional claims in published maps and institutional affiliations.

Cover illustration: Getty/Jose A. Bernat Bacete

Printed on acid-free paper

This Palgrave Macmillan imprint is published by Springer Nature
The registered company is Springer International Publishing AG
The registered company address is: Gewerbestrasse 11, 6330 Cham, Switzerland

*This book is dedicated to my parents, Thomas and Dorothy Pink,
with love and gratitude*

Acknowledgments

The research and writing of this book on climate change has been a fascinating and rewarding journey of learning. Many people have contributed to its completion. I should like to express my sincere appreciation to the following generous and knowledgeable people: Rachael Ballard and the editorial team at Palgrave Macmillan, Premier Peter Taptuna, Dr. Alan Davis, The Honorable Thomas Mulcair, Isabel Alvarez, Dr. Apichart, Niall O'Connor, Dr. Shew-Jiuan SJ Su, Dr. Jiun-Chuan Lin, Zamlha Tempa Gyaltsen, Dr. Tingju Zhu, Dr. Anne Nyatichi Omambia, Dr. Moctar Dembele, Dr. Jean Rasasolofoniaina, Dr. Allen Swagoto Baroi, Dr. Mizanur Rahman, Dr. Khaled Hassen, Dr. Mehmood Ul-Hassan, Dr. Heiko Balzter, Mayor Eva Pinnerod, Dr. Greg Millard, Dr. Diane Purvey, Dr. Shinder Purewal, Dr. Francis Abiew, Dr. Noemi Gal-Or and my wonderful family.

Contents

1 Introduction — 1

2 North America: Canada and the USA — 15

3 South America: Brazil, Ecuador, Argentina — 51

4 Southeast Asia: Thailand, Myanmar, Japan — 75

5 China — 109

6 Africa: Kenya, South Africa, Botswana — 125

7 India — 163

8 The Middle East: Egypt, Israel, Jordan — 185

9 Europe: UK, Italy, Greece — 217

Conclusion: Climate Change Projections to 2100   257

Appendix A: The Paris Agreement   261

Appendix B: Climate Change Glossary   293

Index   295

1

Introduction

"Necessity ... the mother of invention."
Plato

Climate change is a worldwide phenomenon with profound impacts on human development and the global environment. These effects are decisively demonstrated by rising sea level, increased flooding, saltwater intrusion, drought, extreme weather patterns, increasing health emergencies particularly in developing countries, crop destruction, skyrocketing food prices and severe water scarcity projected to affect 2 billion people by 2050.

According to the United Nations Intergovernmental Panel on Climate Change (UNIPCC) 4th Assessment Report, from 2020 to 2050 the global average temperature will rise by 2.6 °C, sea level rise will be at least 1 m (10% of global population live in coastal areas), and crop yields will decline significantly and by 50% in some African countries by 2020.[1] The report warned of sea level rise of 18–59 cm by 2100, which is regarded as a conservative estimate. Climate scientists also warn of no Arctic ice by 2070. Combined, these effects will be profound and have alarming consequences for the natural world, including human life.

Every year the deaths of more than 300,000 people can be attributed to climate change. It seriously affects a further 325 million people and causes annual economic losses of USD 125 billion.[1] Four billion people are vulnerable to the effects of climate change and 500–600 million people—around 10% of the planet's human population—are at extreme risk. As such, climate change has been recognized as a fundamental threat to human rights.[2]

Another sobering reality is the emergence of climate change refugees numbering hundreds of millions. The respected IPCC has adopted the research findings of Prof. Norman Myers, which state a figure of at least 200 million climate change refugees by 2050. Although controversial, this figure and its daunting implications are by no means the final word. Other scholars and climate scientists have calculated numbers as high as 500 million, including Dr. Mehmood Ul Hassan, Director of Capacity Development at the World Agroforestry Center. Europe is struggling to accept and absorb fewer than 1 million Syrian refugees, in stark contrast to the humanitarian openness of Turkey and Jordan. One wonders how the world community will respond to the coming global climate change refugee crisis. Moreover, the Conference of Parties (COP21) that took place in Paris in December 2015, although notable for adopting a new and stringent global approach to carbon emissions, alarmingly omitted any reference to climate refugees in its final communiqué, thus completely ignoring an issue that demands urgent and substantive action. According to Marine Franck, coordinator of the Advisory Group on Climate Change and Human Mobility, "Climate-related displacement is not a future phenomenon. It is a reality; it is already a global concern."[3]

Anthropogenic (man-made) greenhouse gas (GHG) emissions have increased dramatically since the Industrial Revolution. These gases include carbon dioxide (CO_2), methane and nitrous oxide, and are the main cause of global warming. "About 40% stays in the atmosphere, 60% goes into the oceans, soil, trees and plants. Scientists note that an alarming 50% of CO_2 emissions since 1750 occurred in the period 1975–2015."[4] After a decade of very high annual growth rates of global CO_2 emissions of 4% on average, the growth in emissions almost stalled in 2014 with an increase of only 0.5% to 35.7 billion tons (Gt) CO_2, with the largest emitting countries in that year being China—30%; the USA—15%; the European Union—9.6%; India—6.6%; Russia—5.0%; and Japan—3.6%.[5] In 2013, 39.8 billion tons (36.1 billion metric tons) of CO_2 was emitted into the environment from the burning of oil, gas and coal. This figure represents a 778 million ton (706 metric tons) or 2.3% increase from 2012.

The fact that 90% of climate change is anthropogenic is a source of both despair and hope. Despair because human action and governmental indifference have escalated environmental destruction and intensified human suffering. Hope because human action at international, national, regional and local levels has the capacity to respond with innovative and responsible policies that promote, for example, rainwater harvesting, transition towns, electric vehicle adoption, wind turbines, solar energy, cloud seeding and desalinization to mitigate some of the destructive impacts of climate change. The World Bank (2010) estimated that annual investments of USD 70–100 billion will be needed for effective mitigation strategies and that the annual destructive cost of climate change to be between USD 700 billion and USD 1 trillion.

The United Nations Framework Convention on Climate Change (UNFCCC), in Article 1, defines climate change as, "a change of climate which is attributed directly or indirectly to human activity that alters the composition of the global atmosphere and which is in addition to natural climate variability observed over comparable time periods."

There are several critical concerns related to climate change that have been noted by authoritative scientific climate studies.

> **Hazard:** The potential occurrence of a natural or human-induced physical event or trend or physical impact that may cause loss of life, injury, or other health impacts, as well as damage and loss to property, infrastructure, livelihoods, service provision, ecosystems and environmental resources.
>
> **Exposure:** The presence of people, livelihoods, species or ecosystems, environmental functions, services and resources, infrastructure, or economic, social, or cultural assets in places and settings that could be adversely affected.

Vulnerability: The propensity or predisposition to be adversely affected. Vulnerability encompasses a variety of concepts and elements including sensitivity or susceptibility to harm and lack of capacity to cope and adapt.

Impacts: Effects on natural and human systems. The impacts of climate change on geophysical systems, including floods, droughts and sea level rise, are a subset of impacts called physical impacts.

Risk: The potential for consequences where something of value is at stake and where the outcome is uncertain, recognizing the diversity of values. Risk is often represented as probability of occurrence of hazardous events or trends multiplied by the impacts if these events or trends occur.

Adaption: The process of adjustment to actual or expected climate and its effects. In human systems, adaption seeks to moderate or avoid harm or exploit beneficial opportunities.

Resilience: The capacity of social, economic and environmental systems to cope with a hazardous event or trend or disturbance, responding or reorganizing in ways that maintain their essential function, identity and structure, while also maintaining the capacity for adaption, learning, and transformation.[6]

Although climate change is recorded through the millennia, the incidence and speed of occurrence has climate scientists concerned and citizens from Africa to the Arctic experiencing its alarming effects. It is evident that the modern age of climate change began with the Industrial Revolution. "In the period 1750 to 2011, cumulative anthropogenic CO_2 emissions to the atmosphere were 2040 ± 310 $GtCO_2$ and each successive decade since 1850 has witnessed successively warmer weather. Moreover, the period from 1983 to 2012 was likely the warmest 30-year period of the last 1400 years in the Northern Hemisphere, where such assessment is possible."[7]

There is also a noticeable trend toward ocean pollution, warming and acidification. As a result of elevated CO_2 in the atmosphere the world's oceans are 30% more acidic today than before the Industrial Revolution. Indeed, the World Wildlife Fund has noted damage to fish stocks globally that may culminate in the complete depletion of fish stocks by 2070. In the past 250 years, the oceans worldwide have subsumed approximately 560 billion tons of CO_2, which accounts for the rapid rate of increase in ocean acidity. Another fascinating finding is the impact of climate change on solar irradiance. The prime source of energy to the Earth is radiant energy from the Sun. Radiant energy is measured as solar irradiance. Scientists calculate that the release of GHGs from fossil fuel burning is causing a growing percentage of outgoing thermal radiation coming from the Earth. An IPCC report noted that, "The resulting imbalance between incoming solar radiation and outgoing thermal radiation will likely cause the Earth to heat up over the next century, possibly melting polar ice caps, causing sea levels to rise, creating violent global weather patterns, and increasing vegetation density."[8]

> Since the beginning of the industrial era, oceanic uptake of CO_2 has resulted in acidification of the ocean; the pH of ocean surface water has decreased by 0.1 (high confidence), corresponding to a 26% increase in acidity, measured as hydrogen ion concentration.[9]

The depletion of Arctic ice levels and related global sea level rise are phenomena that will have consequential impacts upon human and marine life. Overall, coastal flooding threatens approximately 10% of global population who live in coastal communities. As sea level continues to rise, these communities will increasingly be threatened by flooding, saltwater intrusion, disease and the destruction of land and property. One dramatic example is the Marshall Islands, a peaceful island nation of 53,000 people. The long-term forecast for the Marshall Islands is daunting, with a projection of total water submersion by 2100. The US government has already settled more than 7000 Marshallese citizens in Arkansas and pledged to resettle the entire remaining population when the situation becomes urgent. Severe flooding projections are also indicated for the populous South Asian nation of Bangladesh whose 164 million population (2017) is expected to reach 202 million by 2050. Climate scientists warn of a high volume of climate change refugees from Bangladesh due to the extreme vulnerability of the country to coastal and internal flooding. Recent studies show that due to the effect of sea level rise the densely populated coastal zone of Bangladesh is becoming highly vulnerable to coastal floods; whereas glacier melts in the Himalayan region cause flash floods in the mountainous regions and the foothills of Nepal and this extends to the northern region of Bangladesh.[10]

"The annual mean Arctic sea-ice extent decreased over the period 1979 to 2012, with a rate that was very likely in the range 3.5 to 4.1% per decade. Arctic sea-ice extent has decreased in every season and in every successive decade since 1979, with the most rapid decrease in decadal mean extent in summer."[11] Over the period 1901 to 2010, global mean sea level rose by 0.19 (0.17 to 0.21) m. The rate of sea level rise since the mid-nineteenth century has been larger than the mean rate during the previous two millennia.[12] While global sea level rose about 17 cm in the twentieth century, the rise from 2007 to 2017 alone has been nearly double that level.

A recent study projected three global climate change scenarios to 2050. The studies reported on projections that were (a) expected, (b) severe and (c) catastrophic.

Expected climate change outcomes include the following.

1. By 2040, the average global temperature rises by 1.3 °C (2.3 °F) above 1990 levels.
2. Global sea level rises by 0.23 m (0.75 ft) causing damage to the most vulnerable coastal wetlands and negative impacts on local fisheries.
3. The most significant impact of climate change affects southwestern USA; Central America; sub-Saharan Africa; the Mediterranean region; the mega deltas of South and East Asia; the tropical Andes; and small tropical islands of the Pacific and Indian oceans.
4. Many of the affected areas have large, vulnerable populations requiring international assistance to cope with or escape the effects of sea level rise.[13]

Severe climate change outcomes include the following.

1. By 2040, the average global temperature rises 2.6 °C (4.7 °F) above 1990 levels.
2. Global sea level rises by 0.52 m (1.7 ft).
3. Water availability decreases significantly in the most affected regions at lower altitudes (dry tropics and sub-tropics) affecting 2 billion people.

4. Developing nations at lower altitudes are affected most severely because of climate sensitivity and low adaptive capacity. Industrialized nations to the north experience clear net harm and must divert greater proportions of their wealth to combat climate change at home.[14]

Catastrophic climate change outcomes include the following.

1. Between 2014 and 2100, the average global temperature rises 5.6 °C (10.1 °F) above 1990 levels.
2. Global sea level rises by 2 m (6.6 ft).
3. The North Atlantic Meridional Overturning Circulation (MOC) collapses mid-century generating large-scale disruption of North Atlantic marine ecosystems and associated fisheries.
4. Water availability and loss of food security disproportionately affect poor countries at lower altitudes. Extreme weather events are more or less evenly distributed.[15]

The severe human rights and environmental challenges of climate change can be linked to the new paradigm of human security first enunciated by the United Nations Development Program (UNDP) in a landmark 1994 report. Human security is a new approach to human rights that rejects the state centered "realist" perspective that posits state interests are the first and only consideration of a government. In contrast, human security favors a "people centered" approach that protects citizens from seven basic threats: community, environmental, economic, food, health, personal and political. The increasing inability of the Indian government to meet the basic water needs of its citizens; the detection in recent years of "cancer villages" in China with skyrocketing cancer rates (well above national averages) caused by the deliberate dumping of toxins in village lands and waters for several decades; and the struggle of the US government and state government of Louisiana to rebuild and support struggling impoverished communities after the devastating impact of Hurricane Katrina are clear examples of human security threats.

According to the human security paradigm, these seven threats confront individuals in society in situations of war, extreme poverty, marginalization and deprivation. The goal of human security is to focus on the individual in the context of minimizing and ideally removing one or more of these threats. Climate change and water insecurity pose enormous threats to the individual in each of the seven areas highlighted by human security. Accordingly, human security has the potential to be a powerful and important guideline at international, national and community level to support citizens facing the daunting and increasing problems of climate change.

Increasingly, the military and national security apparatus in many countries are sounding the alarm over the effects of climate change on national and international security. In the USA, military planners and experts are also raising concerns. The US Department of Defense in its 2015 report to Congress, "National Security Implications of Climate Related Risks and a Changing Climate," warned of several troubling outcomes, "Global climate change will have wide-ranging implications for U.S. national security interests over the foreseeable future because it will aggravate existing problems—such as poverty, social tensions, environmental degradation, ineffectual leadership, and weak political institutions—that threaten domestic stability in a number of countries."[16]

Clearly, the ability of the US government and society to benefit from global engagement will be limited and in some cases threatened by the growing incidence of disruption and crisis caused by climate change. As the report notes,

> DoD [Department of Defense] recognizes the reality of climate change and the significant risk it poses to U.S. interests globally. These impacts are already occurring, and the scope, scale, and intensity of these impacts are projected to increase over time, and include: (1) persistently recurring conditions such as flooding, drought, and higher temperatures increase the strain on fragile states and vulnerable populations by dampening economic activity and burdening public health through loss of agriculture and electricity production, the change in known infectious disease patterns and the rise of new ones, and increases in respiratory and cardiovascular diseases; (2) more frequent and/or more severe extreme weather events that may require substantial involvement of DoD units, personnel, and assets in humanitarian assistance and disaster relief (HA/DR) abroad and in Defense

Support of Civil Authorities (DSCA) at home; (3) sea level rise and temperature changes lead to greater chance of flooding in coastal communities and increase adverse impacts to navigation safety, damages to port facilities and cooperative security locations, and displaced populations and (4) decreases in Arctic ice cover, type, and thickness will lead to greater access for tourism, shipping, resource exploration and extraction, and military activities."[17]

Thus there are enormous human security and national security implications to climate change as well as the environmental impacts. One of the tragic outcomes is the impact on poor and developing countries. The IPCC and numerous studies have reported the disproportionate burdens that will fall on marginalized populations and developing countries. For example, Laos, the poorest country in Southeast Asia, has recently achieved substantial progress in the battle against malaria. In Laos, the average temperature is projected to rise by 4.6 °C in a high emission eventuality and the country is expected to experience increased flooding and rainfall episodes. These factors in a tropical country will aggravate the mosquito problem thus propelling chronic and more serious outbreaks of malaria. According to the World Health Organization, malaria causes approximately 420,000 fatalities each year. These figures may spike considerably under climate change scenarios that indicate increased flooding and extreme heat.

Approximately 1.4 billion global citizens live on less than USD 1.25 per day, which is below the World Bank measurement of extreme poverty. Moreover as Nobel Laureate Amartya Sen has noted, capability deprivation affects the poor in a myriad of ways that cannot be measured by income calculations alone. Extreme poverty affects health, mobility, education, gender equality, the attainment of basic amenities, opportunities for development and the ability to adapt to changes in the environment. Overall, due to serious economic, health, infrastructure and social impediments in numerous developing nations the ability to respond to the negative impacts of climate change will be limited. A report by the Brookings Institute offered a sober note: "We already know that such climate effects as sea-level rise, droughts, heat waves, floods, and rainfall variation could, by the 2080s, push another 600 million people into acute malnutrition, increase the number of people facing water scarcity by 1.8 billion, and increase those facing coastal flooding by many millions."[18] In Africa, the projections are dire for many regions in an escalating spiral of climate related stresses: "Already, roughly a quarter of Africa's population is under high-water stress; by 2020, the population at risk is projected to be 75 to 250 million people. By the 2080s, Africa's arid and semi-arid terrain may expand by 5 to 8 percent, and its wheat production may cease entirely. Sea-level rise will imperil coastal areas. Malaria will spread."[19]

Efforts to address one of the core concerns of climate change—global warming—have been spearheaded through the important work of the UNFCCC. In 1992, a shaft of light cut into the darkness of environmental degradation when countries banded together to sign an international treaty, the UNFCCC, that formed a blueprint to combat climate change by limiting average global temperature increases that contribute to global warming and to address their impact. In 1997, the Kyoto Protocol was adopted, which legally binds developed country Parties to emission reduction targets. The Kyoto Protocol's first commitment period started in 2008 and ended in 2012. The second commitment period began on January 1, 2013, and will end in 2020. There are 197 Parties to the Convention and 192 Parties to the Kyoto Protocol. The 2015 Paris Agreement, adopted in Paris on December 12, 2015, was a landmark agreement to move forward in the battle against climate change global warming by limiting global temperature rise in the twenty-first century to below 2 °C above pre-industrial levels and to pursue efforts to limit the temperature increase further to 1.5 °C. To date, 132 Parties have ratified the Agreement of the 197 Parties to the Convention. On October 5, 2016, the threshold for entry into force of the Paris Agreement was achieved when 132 Parties ratified the Agreement of the 197 Parties to the Convention. The Paris Agreement entered into force on November 4, 2016.

The global impact of climate change will affect many regions with devastating effect. The countries most at risk of drought are projected to be Malawi, Ethiopia, Zimbabwe, India and Northern China, with more than 300 towns and villages experiencing the effects of desertification.

Malawi is a low-income nation with the majority of citizens living in rural areas where the median income is approximately USD 975 per year. There have been two severe droughts in the country from 1995 to 2015. Major flooding will affect Bangladesh, China, India and Cambodia. In Bangladesh, 30–70% of the country is flooded each year as waters move from the Himalayan Ranges to the Bay of Bengal in the south, and climate change is expected to exacerbate this problem. Coastal flooding will increasingly impact agriculture and the livelihood of millions of citizens. Extensive storms are expected to severely impact Bangladesh, the Philippines, Madagascar and Vietnam. Yet all continents and countries of the world will experience the effects of climate change in varying degrees of intensity. These challenges will test the resolve, ingenuity and financial resources of every government and society.

Notes

1. IPCC, 2007: Climate Change 2007: The Physical Science Basis. Contribution of Working Group I to the Fourth Assessment Report of the Intergovernmental Panel on Climate Change [Solomon, S., D. Qin, M. Manning, Z. Chen, M. Marquis, K.B. Averyt, M.Tignor and H.L. Miller (eds.)]. Cambridge University Press, Cambridge, United Kingdom and New York, NY, USA.
2. EJF (2009) No Place Like Home—Where next for climate refugees? Environmental Justice Foundation: London.
3. Marine Franck, the coordinator of the Advisory Group on Climate Change and Human Mobility. COP21 Press Conference, December 2015.
4. IPCC 4th Assessment Report.
5. Trends in global CO_2 emissions: 2015 Report © PBL Netherlands Environmental Assessment Agency The Hague, 2015.
6. IPCC, 2014: Summary for policymakers. In: Climate Change 2014: Impacts, Adaptation, and Vulnerability. Part A: Global and Sectoral Aspects. Contribution of Working Group II to the Fifth Assessment Report of the Intergovernmental Panel on Climate Change [Field, C.B., V.R. Barros, D.J. Dokken, K.J. Mach, M.D. Mastrandrea, T.E. Bilir, M. Chatterjee, K.L. Ebi, Y.O. Estrada, R.C. Genova, B. Girma,

E.S. Kissel, A.N. Levy, S. MacCracken, P.R. Mastrandrea, and L.L. White (eds.)]. Cambridge University Press, Cambridge, United Kingdom and New York, NY, USA, pp. 1–32.
7. Ibid.
8. IPCC, 2001.
9. IPCC, 2014: Summary for policymakers. In: Climate Change 2014: Impacts, Adaptation, and Vulnerability. Part A: Global and Sectoral Aspects. Contribution of Working Group II to the Fifth Assessment Report of the Intergovernmental Panel on Climate Change [Field, C.B., V.R. Barros, D.J. Dokken, K.J. Mach, M.D. Mastrandrea, T.E. Bilir, M. Chatterjee, K.L. Ebi, Y.O. Estrada, R.C. Genova, B. Girma, E.S. Kissel, A.N. Levy, S. MacCracken, P.R. Mastrandrea, and L.L. White (eds.)]. Cambridge University Press, Cambridge, United Kingdom and New York, NY, USA, pp. 1–32.
10. Dewan, Tanvir H., "Societal impacts and vulnerability to floods in Bangladesh and Nepal" in Weather and Climate Extremes 7 (2015), pp. 36–42.
11. IPCC, 2014: Summary for policymakers. In: Climate Change 2014: Impacts, Adaptation, and Vulnerability. Part A: Global and Sectoral Aspects. Contribution of Working Group II to the Fifth Assessment Report of the Intergovernmental Panel on Climate Change [Field, C.B., V.R. Barros, D.J. Dokken, K.J. Mach, M.D. Mastrandrea, T.E. Bilir, M. Chatterjee, K.L. Ebi, Y.O. Estrada, R.C. Genova, B. Girma, E.S. Kissel, A.N. Levy, S. MacCracken, P.R. Mastrandrea, and L.L. White (eds.)]. Cambridge University Press, Cambridge, United Kingdom and New York, NY, USA, pp. 1–32.
12. Ibid.
13. Gulledge, Jay. "Three Plausible Scenarios of Future Climate Change." Climatic Cataclysm: The Foreign Policy and National Security Implications of Climate Change, edited by Kurt M. Campbell, Brookings Institution Press, 2008, pp. 49–96, www.jstor.org/stable/10.7864/j.ctt1262fp.6
14. Ibid.
15. Ibid.
16. "National Security Implications of Climate Related Risks and a Changing Climate," United States Department of Defense, 23 July 2015.
17. Ibid.

18. United Nations Development Program, Human Development Report 2007/2008: Fighting Climate Change—Human Solidarity in a Divided World (New York: United Nations Development Program, 2007). (http://hdr.undp.org/en/media/HDR_20072008_EN_Complete.pdf [December 2008]), pp. 27–30.
19. M. Boko, I. Niang, A. Nyong, C. Vogel, A. Githeko, M. Medany, B. Osman-Elasha, R. Tabo, and P. Yanda, "2007: Africa," in Climate Change 2007: Impacts, Adaptation and Vulnerability—Contribution of Working Group II to the Fourth Assessment Report of the Intergovernmental Panel on Climate Change, ed. M.M. Parry, O.F. Canziani, J.P. Palutikof, P.J. van der Linden, and C.E. Hanson (Cambridge UP, 2007), pp. 433–467.

2

North America: Canada and the USA

North America is the third largest continent in the world and located in the northern hemisphere. It is distinguished by vast resources, extreme weather patterns, 23 nations and a population of 585 million citizens reaching from Nunavut to Nicaragua. In terms of global economic position, the USA is first, Canada is tenth and Mexico is 15th. Its total land mass is 9.5 million square miles, which represents 16.5% of global land mass. There are several oceans including the Arctic Ocean in the north, the Pacific Ocean and the Atlantic Ocean. The Caribbean Sea rests on the southern boundary. In the far north, the Canadian Arctic is composed of an astonishing 36,563 islands, which helps to give Canada the longest coastline in the world. Despite its relative wealth, North America is a region that will face considerable climate change effects that will continue to mount over the coming decades. The most notable features will be drought, flooding, extreme weather patterns, extreme heat, rising health crises and severe air pollution in major cities such as Toronto, New York, Los Angeles and Mexico City.

Climate scientists have noted predictions for the continent with a high degree of certainty that include: wildfire-induced loss of ecosystem integrity, property loss, human morbidity and mortality as a result of increased

© The Author(s) 2018
R. M. Pink, *The Climate Change Crisis*,
https://doi.org/10.1007/978-3-319-71033-4_2

drying trends and temperature trends, heat related human mortality in the southern USA and Mexico, social system disruption, public health impacts, water-quality impairment due to sea level rise, extreme precipitation and cyclones, higher sea levels and associated storm surges, more intense droughts and increased precipitation variability—projected to lead to increased stresses to water, agriculture, economic activities and urban and rural settlements.[1] Although climate refugees are expected in their millions in the developing world, which will bear the harshest impact of climate change, many people will be surprised to learn that climate refugees will be seen in their millions in North America. For example, expected severe flood episodes in the US coastal cities of New York, New Orleans, Houston and Los Angeles may generate up to 42 million refuges by 2050. The devastation of Hurricane Katrina in 2005 killed an estimated 1833 people, left millions homeless along the Gulf Coast, and in New Orleans caused USD 108 billion in property damage, rendering it the costliest storm in US history. Severe and unpredictable weather patterns are one of the consequences of climate change and are expected to increase in intensity in North America and indeed in all regions of the world.

Canada

> "The beauty of the trees, the softness of the air, the fragrance of the grass, speaks to me."
> Chief Dan George

Canada is the 11th largest economy in the world and has abundant natural resources. In 2016, the country ranked tenth in world gross domestic product (GDP) at USD 1.53 trillion and ninth in the United Nations (UN) Human Development Index. It produces the third largest amount of hydroelectric power in the world and has the fifth largest oil reserves, is the fifth largest coal producer and fifth largest dry natural gas producer. Canada has the largest coastline in the world at 202,080 km (125,567 miles) of coastline, the largest water area and is ranked as one of the top five water rich nations. It also contains the largest forest cover of any nation at 4,916,438 km^2, which is 10% of the global total. With a small, relatively educated population of 35 million it is well positioned to manage and respond adaptively to the significant challenges of climate change.

In 2011, CO_2 contributed 79% of Canada's total GHG emissions. In 2015, Canada was the ninth largest global CO_2 emitter at 1.71% of total emissions. The majority of these emissions are produced by the combustion of fossil fuels. Methane (CH_4) accounted for 13% of Canada's total emissions, largely from fugitive emissions from oil and natural gas systems, as well as activities in the agriculture and waste sectors. In 2030, Canada's emissions are projected to be 815 Mt CO_2 equivalent, or 11% above 2005 levels, with current measures in place.[2] In 2012, Canada's share of global GHG emissions was 1.6%. The largest emitters in order of volume are China, the USA, the European Union (EU), India and Russia. Canada ranks 11th. Total GHGs have increased approximately 47% since 1990. This dramatic rise from 1980 to 2018 alone underscores the validity and importance of the 2015 COP21 summit in Paris designed to limit global warming and the associated trigger of GHGs. In the historic 2015 Paris Agreement, signatory governments agreed to limit the rise in average world temperature to well below 2 °C (3.6 °F) above pre-industrial times with a target of 1.5 °C.

Between 1948 and 2012, the annual average surface air temperature over Canada's landmass increased by about 1.7 °C, approximately twice the global average. Stronger trends are found in the north and west, particularly during the winter and spring. Northern Canada, north of the 60° latitude, has increased in temperature at a rate of approximately 2.5 times the global average since the late 1940s.[3] It is well documented that Canada's Arctic will experience monumental geological and terrain shifts with the dramatic sea ice melt that is predicted. Most scientific observers calculate that Arctic ice will disappear by 2050. The impact on human populations, such as the citizens of the Canadian territory of Nunavut, on the economy, animals, the ecosystem, marine life and the environment will be immensely disruptive. The melting of permafrost alone will cost billions of dollars in infrastructure damage and replacement costs.

Flooding and Drought

Major climate change effects on Canada include: "reduced arctic ice cover, changes in timing and amount of surface water availability, increased evaporation contributing to lower levels in the Great Lakes, increased depth and extent of permafrost thaw, decreased quality and shorter seasons for northern ice roads, increased loss of forests due to pests and wildfires, more frequent droughts and flooding, and increased risks from food-borne diseases."[4] Further problems likely to affect Canada include increased annual precipitation, drier summers particularly in Manitoba, Ontario and Quebec, decreasing snow levels yet higher snow volumes in northern Canada.

Flooding causes mass property damage and significant community upheaval in Canada. The projections therefore of elevated flood risks due to climate change are cause for serious concern among citizens, communities and policymakers at all levels of government. Recent studies have enumerated the comprehensive nature of flood damage, which affects millions of citizens and multiple sectors of the economy. These impacts include: property damage, (KPMG 2014; Oulahen 2014; Public Safety Canada 2015d); population displacement (Levine, Esnard and Sapat 2007); destruction of infrastructure (Kidd 2011); damage and interruption to business operations (Ingirige and Wedawatta 2011); employment displacement (Davies 2016); threats to physical health (Burton et al. 2016; Carroll et al. 2010); and mental health maladies, including post-traumatic stress disorder, depression and anxiety (Lamond, Joseph and Proverbs 2015; Stanke et al. 2012). For example, severe flooding in Calgary in 2013 caused the most costly natural disaster in Canadian history with thousands of people displaced and an estimated USD 6 billion in property and infrastructure damage.

A recent report concludes that the social and economic costs associated with flooding are increasingly becoming unsustainable for Canadian society. "First, there has been a dramatic increase in urban flood damages over the past two decades, second, the risk of urban flood damage is increasing, third, public expenditures on disaster financial assistance programs have expanded dramatically and a final indicator that flood risk is no longer socially acceptable is the attempts by affected local residents to seek restitution through the courts."[5]

Drought has been an historic issue in the Prairie Provinces and interior of BC. Indeed, the town of Osoyoos, BC, the only dessert area in Canada, has recorded temperatures exceeding 40 °C. "Evidence suggests that the incidences of drought increased over most of the country between 1950 and 2002. Several long duration and severe droughts have affected the southern Canadian Prairies and interior valleys of British Columbia over the last century. Available studies mainly suggest that future droughts will likely be more frequent, longer lasting and more severe in those regions that already experience these events (*i.e.*, more southern and interior areas of the country)."[6]

Extreme Weather

A 2017 report highlighting climate change impacts on Canadian municipalities noted multiple threats. "Climate change poses significant risks for Canadian municipalities, particularly in the form of extreme weather, such as severe thunderstorms, ice storms, hailstorms, windstorms, blizzards, and tornadoes. Canada's major cities are especially vulnerable to extreme weather, due to their large, dense populations, valuable and geographically concentrated property, and complex, interdependent infrastructure networks, all of which are susceptible to threats from localized climate hazards."[7] The ferocity of nature and extreme weather unpredictability were destructively demonstrated in Toronto, the country's largest city, when a sudden and dramatic weather storm in 2013 caused major flooding affecting 300,000 citizens and causing USD 940 million in property damage.

Several recent examples of extreme weather events have buffeted Canada resulting in substantial human suffering and property damage. According to an Environment Canada report, "Economic losses from such events in Canada are often in the hundreds of millions of dollars (e.g. Hurricane Juan, Alberta hailstorms, British Columbia wildfires), and even in the billions (1998 Ice Storm, 1996 Saguenay flood; 2001–2002 national-scale drought). Insect damage to forests and crops may also be significant."[8]

A report for the Insurance Bureau of Canada offers a sober assessment of the major climate change implications including extreme weather events that will beset the Canadian population. Notable impacts include:

1. Higher temperatures will lead to a dramatic increase in wildfires—50–500%, particularly over northwestern Ontario.
2. The cost of natural catastrophic losses have been more than USD 1 billion and will likely continue—since the 1980s, losses have been doubling every five to ten years. Another indicator of the impact of the changing climate is that water damage has now surpassed fire as the most frequent cause of insurance claims in Canada.
3. Governments need to look at replacing aging sewer systems that won't be able to handle the intense rainfalls, and homeowners should ensure they have backstops on their drains and that their properties are properly graded.
4. The number of frost-free days is expected to double.
5. The number of days with temperatures exceeding 30 °C will increase in Toronto from about 15 in 2005 to 28 in 2050.
6. Winter precipitation will increase by 20% near Hudson Bay and by 10% in the south.
7. Summer precipitation will jump by about 5% in the north with a smaller increase in the south.
8. Freezing rain events lasting six hours a day or longer will rise by 35% in southwestern Ontario and by 80% in the east. In the north, the jump will be 80–100%.[9]

Health, Human Development and Food Security

Even a prosperous and highly developed country such as Canada will experience notable impacts upon health and human development due to climate change. Vulnerable groups, including the elderly, children, health challenged citizens and vulnerable climate communities are expected to experience increasing levels of difficulty. A government of Canada report states that "Stronger evidence has emerged since 2008 of the wide range of health risks to Canadians posed by a changing climate. For example, climate-sensitive diseases (e.g. Lyme disease) and vectors are moving northward into Canada and will likely continue to expand their range. In addition, new research suggests climate change will exacerbate air pollution issues in some parts of Canada."[10]

Impact studies on the correlation between climate change and heath vulnerabilities in Canada have enumerated a wide range of concerns that will increasingly demand attention. These include:

1. more frequent, severe and longer heat waves;
2. warmer weather, with possible colder conditions in some locations;
3. heat related illnesses and deaths;
4. respiratory and cardiovascular disorders;
5. possible changed patterns of illness and death due to cold temperatures;
6. death, injury and illness from violent storms, floods and so on;
7. psychological health effects, including mental health and stress related illnesses;
8. health impact of food or water shortages;
9. illnesses related to drinking-water contamination;
10. effects of the displacement of populations and crowding in emergency shelters;
11. indirect health impact of ecological changes, infrastructure damage and interruptions to health services;
12. increased air pollution: higher levels of ground-level ozone and airborne particulate matter, including smoke and particulates from wildfires;
13. increased production of pollens and spores by plants;
14. eye, nose and throat irritation, and shortness of breath;
15. exacerbation of respiratory conditions;
16. chronic obstructive pulmonary disease and asthma;
17. exacerbation of allergies;
18. increased risk of cardiovascular diseases (e.g., heart attacks and ischemic heart disease);
19. premature death;
20. increased contamination of drinking and recreational water by runoff from heavy rainfall;

21. changes in the biology and ecology of various disease-carrying insects, ticks and rodents (including their geographical distribution);
22. faster maturation of pathogens within insect and tick vectors;
23. a longer disease transmission season;
24. increased incidence of vector-borne infectious diseases native to Canada (e.g., eastern and western equine encephalitis, Rocky Mountain spotted fever);
25. introduction of infectious diseases new to Canada;
26. possible emergence of new diseases, and re-emergence of those previously eradicated in Canada;
27. depletion of stratospheric ozone by some of the same gases responsible for climate change (e.g., chloro- and fluorocarbons);
28. temperature related changes to stratospheric ozone chemistry, delaying recovery of the ozone hole;
29. increased human exposure to ultraviolet (UV) radiation owing to behavioral changes resulting from a warmer climate;
30. and more cases of sunburn, skin cancers, cataracts and eye damage.[11]

Agriculture is a major sector in Canada accounting for 6.6% of GDP in 2014, USD 101 billion in revenues and 2.3 million jobs in agriculture and food services. Canada is the fifth largest exporter of agriculture and agri-food products after the EU, USA, Brazil and China. There are 64.8 million ha (7%) of Canada's total land area dedicated to agriculture. The serious droughts of 2001 and 2002 and catastrophic flooding in 2010 and 2011 had a harsh impact on citizens, communities and many sectors including agriculture. Extreme weather events can reduce crop yields by as much as 50% during average growing seasons. Thus globally and in Canada the impact of climate change upon land and food systems can be profound. The effects of temperature and climate variability on yields of major crops have been observed with a high probability of occurring, particularly in Ontario and Quebec. Moreover projected increases in temperature, reductions in precipitation in some regions and increased frequency of extreme events would result in net productivity decline in major North American crops by the end of the twenty-first century without significant adaption.[12]

Climate Change Refugees

Although Canada will not be a climate refugee generating country it has experienced temporary severe weather episodes that have displaced thousands of citizens from their homes and caused billions of dollars in property damage. In southern Saskatchewan, historic flooding in 2011 left 4000 citizens temporarily displaced. A storm in May 2013 struck homes at Delta Beach on the south shore of Lake Manitoba. In total, 7100 Manitobans were displaced from their homes, with 2700 still unable to return home by the end of the year. Canada is not at risk of external climate change refugee flows, though it will definitely experience strong migration pressures from other developing countries where climate change refugee numbers are expected to reach into the tens of millions.

Nunavut

Nunavut is an Arctic territorial region of Canada with a population of 33,000 spread over 24 remote communities and a land area of approximately 2 million km^2. It is a territorial government with a legislature and elected premier. Many communities are accessible only by air or boat. With the forecasted disappearance of Arctic ice by the middle of the century and a considerable increase in temperatures and sea level rise, the impact of climate change upon the people, economy and ecosystem of Nunavut will be considerable.

A report by the territorial government of Nunavut stated some of the pronounced climate change effects, which include: "Decreasing sea ice thickness and distribution, which is changing wildlife habitat and affecting and impacting hunters' ability to harvest wildlife; permafrost degradation, changes in ice conditions, rainfall and snow quantity, drainage patterns, temperatures, and extreme weather events can all have implications for existing infrastructure, such as roads and buildings, all of which was designed around a permanently frozen soil regime; increased length of the ice free season may allow for increased shipping through our waterways, including the Northwest Passage; arrival of new insects, birds, fish and mammals previously unknown or rare in Nunavut, and change in the abundance and distribution of familiar animals."[13]

Previous forecasts on the Arctic ice melt ratio appear to be overstated. New research suggests a more rapid cycle of increasing temperatures according to researcher James E. Overland, "Whether a nearly sea ice-free Arctic occurs in the first or second half of the twenty-first century is of great economic, social, and wildlife management interest. It is reasonable to conclude that Arctic sea ice loss is very likely to occur in the first rather than the second half of the twenty-first century, with a possibility of loss within a decade or two."[14] The new meta-study, a study of 36 previous computer models used by the UN and others was utilized to predict Arctic climate change. An ice-free Arctic Ocean is defined as one in which less than 1 million km^2 (386,000 square miles) of ice remain. An additional effect of polar ice melt will be a rise in CO_2 emissions according to one of the world's leading polar experts, Peter Wadhams. According to Wadhams, "By my calculations, the terrestrial warming in the Arctic is roughly equivalent to 25 percent boost in global CO_2 emissions. This, combined with the warming caused by the loss of Arctic sea ice, means that the overall ice/snow albedo effect in the Arctic could add as much as 50 percent to the direct global heating effect of CO_2. But the most worrisome feedback, which could lead to catastrophic effects in the near future, involves the release of seabed methane, a potent greenhouse gas, from the continental shelves of the Arctic Ocean."[15]

There are several significant climate change effects on the people and ecology of Nunavut. A recent report by the government of Nunavut noted the following: "Impacts of varying severity have already been observed in range distribution, habitats, abundance, genetic diversity, behavior and population size in both migratory and non-migratory species. The introduction of new species in Nunavut, including diseases that have not previously occurred in the territory can have a negative effect on existing populations, reduced ice cover, changes to salinity and increased acidity, changes to marine species are likely to result from changes to the ocean ecosystem, as permafrost thaws, it weakens the ground structure, which speeds the process of erosion and slumping, and can cause landslides."[16] Another serious issue facing Nunavut and the Arctic communities in general is the rapid thinning of the ozone layer, which is leading to unhealthy levels of UV rays. The ozone layer protects human, marine and animal life, and plants and crops from dangerous levels of UV radiation.

The thinning of the ozone layer triggers a host of health and environmental problems including elevated risks of skin cancer and cataracts. Since the culture and community of Nunavut has historically lived close to and depended upon the land and nature for sustenance it is therefore particularly vulnerable to the well documented climate change effects on the environment.

Overall, the challenge of climate change mitigation including reductions in GHGs are pressing issues for the Canadian government and society. According to Climate Tracker, Canada is not on target to achieve sustainable targets on these vital issues. "Canada's Nationally Determined Contribution (NDC) is rated 'inadequate.' Under its current policies, Canada will miss both its 2020 pledge and its 2030 NDC targets by a wide margin. Taking into account policies implemented before September 2015, we estimate that Canada's GHG emissions (excluding LULUCF) to increase 23–30% above 1990 levels by 2020. By 2030, emissions are projected to increase by 26–44% above 1990 levels (an increase from 2005 levels by 1–7% and 3–18% in 2020 and 2030 respectively). In October 2016, the new Canadian Government announced a national mandatory carbon-pricing plan that, if enacted, would represent a major step towards policies that could change this adverse outlook. However, more details are required to be able to quantify the impact of the carbon-pricing plan on meeting the NDC goals."[17] In sum, leading global oil production, the continuation of tar sands oil exploitation, a highly industrialized economy and a dominant carbon-based economy are contributing factors that limit effective target achievement for Canada.

Innovation and Adaption Strategies

1. Ratification of the Paris Agreement in 2016.
2. Its NDC communicates its economy-wide target to reduce greenhouse gas (GHG) emissions by 30% below 2005 levels in 2030. After accounting for forestry, we estimate this is a reduction of 20% below 2005 levels and 3% below 1990 levels of GHG emissions, excluding land use, land-use change and forestry
3. (LULUCF). However, Canada is unlikely to meet this target.[18]

4. The Canadian Environmental Protection Act, 1999 (CEPA 1999), directs the government to protect the environment and human health in order to foster sustainable development.
5. Climate and Clean Air Coalition (CCAC) to Reduce Short-Lived Climate Pollutants is established. Since 2012, the CCAC has grown to include more than 70 partners.
6. 2011–2016: The government of Canada allocated USD 148.8 million in funding to support an improved understanding of climate change and to assist Canadians to prepare for the impact of climate change. All provinces in Canada have set up adaption strategies.
7. In 2013 the Canadian government allocated funds for Arctic research through the Natural Sciences and Engineering Research Council's Climate Change and Atmospheric Research initiative. The funding includes USD 32 million over five years to seven university-based research programs.
8. As a result of collective action by governments, consumers and industry, Canada's 2020 emissions are projected to be 128 Mt lower than they would have been under a no-action scenario. This is the equivalent of shutting down 37 coal-fired electricity generation plants. Moreover, Canada's per capita emissions are at a historic low of 20.4 tons of CO_2 equivalent per person—their lowest level since tracking began in 1990. Canada has also demonstrated progress in decoupling emissions growth from economic growth. Since 2005, Canadian GHG emissions have decreased by 4.8%, while the economy has grown by 8.4%.[19]
9. Ratified UNFCCC in 1992.

The USA

"Never does nature say one thing and wisdom another."
President Jimmy Carter

The USA is the largest economy in the world with a GDP of USD 18.5 trillion and ranked eighth in 2015 in the Human Development

Index. In 2015, it was the second largest global CO_2 emitter contributing 15.9% of total emissions. Although the country's wealth will provide a certain measure of protection from climate change and an ability to respond better than most nations, it is still beset by significant climate change challenges that will increase exponentially in the decades to come. The geographical expanse of the USA, stretching from the Arctic to Puerto Rico, leaves the country confronted with multiple and varying climate scenarios from tropical storms in Florida to rapid glacier and permafrost melting in Alaska.

A summary of the major climate change effects in the USA reveals the magnitude of the problem and the tremendous human and financial costs involved. The general trend of increasing temperature has been most pronounced in the northern and western regions of the country. Hurricanes have increased in intensity as well as winter storms. Category four and five hurricanes, the more severe, have increased in frequency. Rising sea levels have been recorded and will show pronounced elevation rates along the Gulf Coast and West Coast. Flooding will increase in intensity and duration particularly in the midwest and northeast, affecting cities such as New York and Boston with devastating effects. Dramatically rising levels of precipitation have been observed from 1960s to 2017, most noticeably in the north and midwest regions. Devastating floods in 1993 destroyed 100,000 homes and caused approximately USD 20 billion in damage along the Missouri and upper Mississippi. In the aftermath, the White House appointed Galloway report discouraged further construction in flood plain areas. One of the clear climate change projections for the USA is more frequent flood episodes, which will jeopardize already vulnerable communities. Increased frequency of heat wave episodes have an impact upon health, water resources, water security, farming capacity and land use. Wildfires are an increasingly dangerous symptom of the rising temperatures associated with climate change in the USA, particularly in the southeast and southwest. Drought and near-drought conditions in several US states, including California, Nevada, Arizona, New Mexico and Texas, will aggravate the wildfire threat and other conditions detrimental to human and socio-economic development.

A scientific report by the National Climate Association (NCA) noted the following eventualities to confront the country:

1. Temperatures in the USA are expected to continue to rise.
2. Increases are also projected for extreme temperature conditions. The temperature of both the hottest day and coldest night of the year are projected to increase.
3. More winter and spring precipitation is projected for the northern USA, and less for the southwest, over the twenty-first century.
4. Increases in the frequency and intensity of extreme precipitation events are projected for all US areas.
5. Short-term (seasonal or shorter) droughts are expected to intensify in most US regions. Longer-term droughts are expected to intensify in large areas of the southwest, the southern Great Plains, and the southeast. Trends in reduced surface and groundwater supplies in many areas are expected to continue, increasing the likelihood of water shortages for many uses.
6. Heat waves are projected to become more intense, and cold waves less intense, everywhere in the USA.
7. Hurricane-associated storm intensity and rainfall rates are projected to increase as the climate continues to warm.[20,21]

Flooding and Drought

Although flood crises have been a constant in US history, their regularity and intensity will be become more pronounced in the future. A report on coastal flooding noted a number of concerns that confront coastal communities in the country. "Events like Superstorm Sandy in 2012 have illustrated that public safety and human well-being become jeopardized by the disruption of crucial lifelines, such as water, energy, and evacuation routes. As climate continues to change, repeated disruption of lives, infrastructure functions, and nationally and internationally important economic activities will pose intolerable burdens on people who are already most vulnerable and aggravate existing impacts on valuable and irreplaceable natural systems."[22]

Sea level rise is another well documented global phenomenon that is aggravated by climate change. In the USA, numerous shorelines are threatened by sea level rise and saltwater intrusion. Saltwater intrusion is a corrosive threat to crops, land and infrastructure. Sea level rise will also have an immense and debilitating impact upon hundreds of coastal communities and major cities such as New York, Boston, Miami, Fort Lauderdale, New Orleans, Houston and Los Angeles. Flooding will damage or destroy infrastructure, cause significant health problems, interrupt water services and place huge demands upon emergency, medical, government and military resources. According to one report, "Assuming that these historical geological forces continue, a 2-foot rise in global sea level by 2100 would result in the following relative sea level rise: 2.3 feet at New York City, 2.9 feet at Hampton Roads, Virginia, 3.5 feet at Galveston, Texas and 1 foot at Neah Bay in Washington state."[23]

In addition to severe flooding, the other devastating climate change certainty to face the USA is intense drought. In 2015, nine US states faced severe to moderate drought conditions. These states included (in order of severity) California, where 93% of the state faced severe drought conditions, Nevada—86%, Oregon—68%, Utah—34%, Arizona—29%, Idaho—26%, Washington—23% and New Mexico—12%. In late 2015, Texas and Oklahoma both faced serious drought conditions. Climate conditions, heat waves and water shortages in some areas are expected to exacerbate the drought situation for these states and potentially other states in the coming years. The regions of the country that will likely face the most proportionate damage from climate change induced weather events and development challenges are the southwest, southeast and northeast.

Southwest

The southwest region includes the states of Arizona, California, Colorado, Nevada, New Mexico and Utah. Their combined population in 2017 was 62 million. Climate change is already having major pronounced effects in the southwest. The populous and prosperous state of California, for example, is now considered a "drought state," a condition that has led to considerable human, financial and resource stress. In the region, temperatures have increased by almost 2 °F in the last century, with the decade 2001–2010 being the warmest since records began 110 years ago.[24] The length of the frost-free season has increased by 19 days in the 1970–2017 period.[25] Average annual temperatures are projected to rise an additional 3.5 °F to 9.5 °F by the end of this century, with the greatest temperature increases expected in the summer and fall.[26] Drought conditions are already common in the southwest, and drought periods are expected to become more frequent, intense and longer, and drought will affect important water sources, including the Colorado River Basin.[27] The steady population growth in the region will leave more and more citizens exposed and vulnerable to a wide range of challenges.

Southeast

The southeast states include Alabama, Arkansas, Louisiana, Florida, Georgia, Kentucky, Mississippi, North Carolina, South Carolina, Tennessee, Virginia and West Virginia with a combined population in 2017 of 89 million. There is a total coastline area of 29,000 miles, which renders the risk of sea level rise, coastal flooding and saltwater intrusion a critical issue of concern. The major cities include Atlanta, Charlotte, Jacksonville, Miami, New Orleans and Orlando. The flood and tropical storm risk for many of the states is severe. The number of days reaching temperatures of greater than 95 °F in the southeast is projected to increase during this century. Since 1970, average annual temperatures in the region have increased by about 2 °F, with the greatest temperature increases occurring during the summer.[28] Additional projections for the region include: temperature increases of 4 to 8 °F by the end of the century; fewer predicted freezing events; inland areas to increase in temperature more than coasts and natural cycles; a substantial increase in the intensity, frequency, duration and strength of Atlantic hurricane activity since the 1980s, among others.[29] New Orleans, Miami and Orlando face increased flood risks and the entire region is subject to severe tropical storms that are anticipated to increase in intensity in the future.

Northeast

The northeast is comprised of nine states including Maine, New Hampshire, Vermont, Massachusetts, Rhode Island, Connecticut, New York, New Jersey and Pennsylvania with a 2017 population of 57 million. The northeast region is also experiencing variable and disruptive weather patterns and significant threats of coastal flooding. States that border the Atlantic Ocean have 2069 miles of coastline and the threat is particularly pronounced for Massachusetts, New York and New Jersey. "Between 1895 and 2011, temperatures rose by almost 2 °F and projections indicate warming of 4.5 °F to 10 °F by the 2080s, the frequency, intensity, and length of heat waves is also expected to increase and the total amount of precipitation and the frequency of heavy precipitation events has also risen in the region."[30] Between 1958 and 2012, the northeast experienced a 70% increase in the amount of rainfall measured during heavy precipitation events, more than any other region in the USA.[31] The cities of Boston and New York in particular face critical coastal flooding and heavy rainfall scenarios.

Extreme Weather

Extreme weather events consist of higher than normal episodes of rainfall, flooding, storms, heat waves, drought conditions that often meet or exceed previous weather related records. There is a strong correlation between elevated climate change symptoms and extreme weather phenomenon. Indeed, Munich Re, the world's largest re-insurance company, documented a list of worldwide disasters between 1980 and 2010 and reached the following conclusions. "In its analysis, 2010 had the second-largest (after 2007) number of recorded natural disasters and the fifth-greatest economic losses."[32] "Although there were far more deaths from geological disasters—almost entirely from the Haiti earthquake—more than 90 percent of all disasters and 65 percent of associated economic damages were weather and climate related (i.e. high winds, flooding, heavy snowfall, heat waves, droughts, wildfires). In all, 874 weather and climate-related disasters resulted in 68,000 deaths and $99 billion in damages worldwide in 2010."[33]

The USA is not insulated from the increased incidents and devastation of extreme weather. Moreover the expected episodes will bring new and deeper challenges to governments and citizens. Some recent examples of extreme weather phenomenon are illustrative of the magnitude of the challenges that are current and expected to increase. Tree ring data suggest that the drought over the last decade in western USA represents the driest conditions in 800 years.[34,35] "The recent heat waves and droughts in Texas (2011) and the Midwest (2012) set records for highest monthly average temperatures, exceeding in some cases records set in the 1930s, including the highest monthly contiguous U.S. temperature on record (July 2012, breaking the July 1936 record) and the hottest summers on record in several states (New Mexico, Texas, Oklahoma, and Louisiana in 2011 and Colorado and Wyoming in 2012). For the spring and summer months, 2012 had the second largest area of record-setting monthly average temperatures, including a 26-state area from Wyoming to the East Coast. The summer (June–August) temperatures of 2012 ranked in the hottest 10% of the 118-year period of record in 28 states covering the Rocky Mountain states, the Great Plains, the Upper Midwest, and the Northeast."[36]

Health, Human Development and Food Security

The impact of climate change upon human health and development are recognized by a wide range of experts to be substantial and sustained for the foreseeable future. The intensity of these effects of course will vary depending upon geographic and climate vulnerability, socio-economic status, existing infrastructure stability, government adaption, response capacity and emergency response. In the USA a number of significant climate change effects are expected with certainty. These include: "Increases in water temperatures, increasing exposure and risk of waterborne illness, changes in exposure to climate or weather-related disasters, rising temperatures and wildfires and decreasing precipitation will lead to increases in ozone and particulate matter, rising sea level and more frequent or intense extreme precipitation, hurricanes, and storm surge events, increased coastal and inland flooding, contaminated water, debris, and disruptions to essential infrastructure, drowning, injuries, mental health consequences, gastrointestinal and other illness."[37]

It is clear that the negative impact of climate change will affect populations disproportionately both internationally as well as domestically. The IPCC has widely reported that more than 85% of the burden of climate change will fall upon people in the developing world. Using a similar comparative matrix, even a relatively prosperous nation like the USA will find disproportionate levels of human suffering and property damage that are determined by a number of factors including age, income, geography, ethnicity, socio-economic status, disability and health.

Research data on the health impacts of climate change emphasize the societal groups who are the most vulnerable. These demographic groups include the following:

1. People who live in poverty may have a difficult time coping with changes. These people have limited financial resources to cope with heat, to relocate or evacuate, or to respond to increases in the cost of food.[38,39]
2. Older adults may be among the least able to cope with the impact of climate change. Elderly people are particularly prone to heat stress.[40]
3. Older residents make up a larger share of the population in warmer areas of the USA. These areas will likely experience higher temperatures, tropical storms or extended droughts in the future.[41] The share of the US population composed of adults over age 65 is also projected to grow from 13% in 2010 to 20% by 2050.[42]
4. Children's growth and development from infancy to adolescence makes them more sensitive to environmental hazards related to climate. For example, because children's lungs develop through adolescence, they are more sensitive to respiratory hazards. Climate change negatively affects air quality because increasing temperatures make it easier for ground-level ozone to form.[43]

Climate change is already having a consequential health impact on the country and is expected to increase the risks. Some of the health conditions that will be exacerbated by climate change include: asthma, chronic obstructive pulmonary disease (COPD), respiratory illnesses, diabetes, cardiovascular disease (CVD), cognitive diseases, obesity and mental illness. By 2030, approximately 41% of the US population is projected to have some form of CVD.[44] Asthma rates in children and obesity across all age groups have increased from 1980 to 2017. Air quality has a major impact upon asthma and respiratory illness rates. Moreover, according to projections from the US government, by 2030 51% of the US population is expected to be obese, and research indicates a 33% increase in obesity and a 130% increase in severe obesity.[45] The number of citizens with a disability will increase. In 2010, the percent of US adults with a disability was approximately 16.6% for those aged 21–64 and 49.8% for persons 65 and older.[46] The number of older adults with activity limitations is expected to grow from 22 million in 2005 to 38 million in 2030.[47]

The USA is one of the largest agricultural producers in the world. In 2016, it was the leading producer of chicken, beef, turkey, maize, soybeans, almonds, strawberries, blueberries and cow's milk. The top ten US agricultural producing states in terms of revenue are California, Iowa, Texas, Nebraska, Minnesota, Illinois, Kansas, North Carolina, Wisconsin and Indiana. All of these states are dramatically affected by episodic extreme weather and in the case of California, Kansas, Texas and Nebraska, serious drought conditions, all of which impact food production, employment and commodity prices. Few industries are as vulnerable to climate change as the agricultural sector. The crops, livestock and seafood produced in the USA contribute more than USD 300 billion to the economy each year.[48] When food-service and other agriculture related industries are included, the agricultural and food sectors contribute more than USD 750 billion to the GDP.[49] Sensitivity and vulnerability to climate change are notable in many aspects of the agriculture sector. For example, "Many weeds, pests, and fungi thrive under warmer temperatures, wetter climates, and increased CO_2 levels. Currently, U.S. farmers spend more than $11 billion per year to fight weeds, which compete with crops for light, water, and nutrients. The ranges and distribution of weeds and pests are likely to increase with climate change."[50]

Climate Change Refugees

Globally, the estimates on climate change refugee numbers range from 200 million (Dr. Norman Myers) to 500 million according to Dr. Mehmood Ul Hassan. In the USA the crisis will be a matter of internal displacement and migration pressure to leave cities and communities after devastating weather events such as hurricanes, coastal flooding, wildfires and severe drought. Researchers at the US Oceanic and Atmospheric Administration assessed sea level change scenarios to the year 2100 for coastal states in high risk areas with moderate to high growth rates. "The data indicates that more than 13 million Americans are at risk with a 6-foot (1.8 meters) rise including 6 million in Florida. With a sea level rise of 3 feet, locations forecast to house 4.2 million people would be at risk of inundation while a doubling of the rise would bring the number to 13.1 million. With densely populated coastal locales, Florida faces the greatest risk, with up to 6.06 million residents projected to be affected if sea levels rise 6 feet, followed by Louisiana (1.29 million people at risk) and California (1 million). Other states that could be heavily impacted in such a scenario include: New York (901,000 at risk), New Jersey (827,000), Virginia (476,000), Massachusetts (428,000), Texas (405,000), South Carolina (374,000) and North Carolina (298,000)."[51]

In 2015, more than 125 million Americans lived in coastal areas, which are already noted as communities vulnerable to climate change flooding and storms. It is expected that four cities, New York, Miami, Houston and Los Angeles, could experience significant climate migration (in the millions) by 2050. Other cities, including Boston, Philadelphia, Atlantic City and Baltimore, also face considerable flooding risk and likely evacuation scenarios involving tens of thousands of citizens.

New York

New York is the largest and most densely populated city in the USA. The 2016 population is 8.6 million. It has a humid, semi-tropical climate, but cold winter weather—the temperature range is 32.6 to 90 °F (0.3 to

32 °C). The city receives 49.9 inches (1270 mm) of rainfall annually. The main natural disasters to affect the city are hurricanes, floods, storms, blizzards, snowstorms and heavy rainfall. As noted in a 2010 City of New York report, the municipal government recognizes the imminent and on-going threats posed by climate change. "We already face climate risks today, including heat waves, blackouts, flooding, and coastal storms. With climate change these risks will only increase."[52]

A NASA report on the climate change implications for New York provided a sober assessment. Specific report projections include the following: "Mean annual temperature has increased a total of 3.4 degrees Fahrenheit (F) from 1900 to 2013. Future mean annual temperatures are projected to increase 4.1 to 5.7 degrees F by the 2050s and 5.3 to 8.8 degrees F by the 2080s, relative to the 1980s base period. The frequency of heat waves is projected to increase from 2 per year in the 1980s to roughly 6 per year by the 2080s. Mean annual precipitation has increased by a total of 8 inches from 1900 to 2013. Future mean annual precipitation is projected to increase 4 to 11 percent by the 2050s and 5 to 13 percent by the 2080s, relative to the 1980s base period. Sea levels have risen in New York City 1.1 feet since 1900. That is almost twice the observed global rate of 0.5 to 0.7 inches per decade over a similar time period. Projections for sea level rise in New York City increase from 11 inches to 21 inches by the 2050s, 18 inches to 39 inches by the 2080s, and, 22 inches to 50 inches, with the worst case of up to six feet, by 2100. Sea level rise projections are relative to the 2000 to 2004 base period."[53]

Innovation and Adaption Strategies

1. The USA ratified the Paris Agreement in 2016
2. For the US "Its Intended Nationally Determined Contribution became its Nationally Determined Contribution (NDC). Under its NDC, the U.S. has pledged to reduce net GHG emissions by 26–28% below 2005 in 2025 including land use, land use change and forestry (LULUCF) (equivalent to 19–24% below 2005 levels excluding LULUCF, and equivalent to 6–12% below 1990 levels excluding

LULUCF). The US's NDC clearly highlights ongoing actions to enhance the regulatory framework, so the achievement of the 2020 and 2025 targets seems feasible."[54]
3. There is a UNFCCC commitment to reduce emissions by 17% by 2020 in relation to 2005 levels.
4. The American Recovery and Reinvestment Act 2009, designates USD 94 billion to renewable energy technology, low carbon vehicles, mass transit and energy efficient initiatives.
5. The Energy Independence and Security Act of 2007 promotes the production of renewable fuels and lowers dependence on oil.
6. The Energy Policy Act 2005 allocates USD 4.3 billion in tax incentives for nuclear power; USD 2.7 billion to extend the renewable electricity production credit; and USD 1.6 billion in tax incentives for investment in clean coal facilities.
7. Former President Jimmy Carter built a new 1.3 MW solar plant on a farm where he once cultivated soybeans and peanuts. The 3852 solar panels, which can move to follow the Sun, can generate enough renewable energy for more than half of the 683-person town of Plains, Georgia.
8. In 1997 the USA signed, but did not ratify the Kyoto Accord.

Interview

Peter Taptuna serves as Premier of Nunavut and has held public office at ministerial and hamlet council levels since Nunavut was established. Proud of his Inuit heritage, culture and strong family ties, Taptuna and his wife live in Kugluktuk. This small community sits on the Coronation Gulf in Canada's Arctic.

While being Premier of Nunavut, what important environmental protection programs has your government initiated or been involved with?

The government of Nunavut has been heavily involved in climate change and environmental protection. In November 2016, the government of Nunavut announced the creation of the Climate Change Secretariat, which is responsible for managing climate change adaption and mitigation programs and policies for the government of Nunavut.

Last year, we also worked with the World Wildlife Fund to host an Arctic Renewable Energy Summit.

In addition, we stood with our fellow provinces and territories in signing the Pan-Canadian Framework for Clean Growth and Climate Change in December 2016. As part of this, we are now working with Canada towards introducing a made-in-Nunavut carbon pricing scheme that recognizes our territory's unique circumstances.

We absolutely need Canada to help invest in clean technology and facilitate adaption and mitigation on multiple fronts, as we do not have the resources.

The environment continues to be a priority for Nunavut, so that Inuit can continue to enjoy traditional ways of life.

What are the most immediate climate change threats facing Nunavut?

Nunavut is at the forefront of climate change. Our weather, our ecosystems, our sea ice and our permafrost are all sensitive to changes in temperature and weather patterns.

Perhaps the most significant threat is the impact that this could have on our way of life and traditional food security. These changes are impacting our ability to access traditional hunting grounds. The hunting and fishing activities that many Nunavummiut use to sustain their way of life are becoming increasingly hazardous, as weather and sea ice patterns become less predictable. As the global temperature rises, the Artic is hit first and hit the hardest.

The UN and leading climate scientists predict that there will be no Arctic glaciers by 2070. What is your assessment of this projection?

One of the most important things when it comes to the Arctic and the circumpolar north with respect to our land, waters and wildlife, is to have science working hand in hand with traditional indigenous knowledge.

There is a tremendous amount of knowledge that our elders possess, which has been passed down from generation to generation. By all accounts, Canada is an Arctic nation with the north comprising about 40% of our land mass. We have the longest jurisdictional coastline in Canada, and Inuit have inhabited this land for millennia. Inuit are very connected to it and have the best understanding of how it has changed over time.

My hope is that science takes a hard look at traditional knowledge and considers this factor thoroughly prior to making any scientific projection. I believe there is a will to better understand the value and significance of traditional knowledge and its role in determining our outcomes. We as indigenous peoples in the circumpolar north expect this collaboration and equal footing with respect to the knowledge we bring to the table.

At a national level, what steps should the Canadian government take to mitigate the impact of climate change in Nunavut and the Arctic?

The Canadian government took an important step by introducing the Pan-Canadian Framework on Clean Growth and Climate Change in December 2016.

Nunavut is not a significant producer of greenhouse gases. In fact, Nunavut contributes less than 0.01% of greenhouse gas, but we directly feel the impacts of climate change associated with emissions from the industrialized world. Signing the framework agreement was an important step, because it brought most of the provinces and territories under one plan and has caveats carved out for the north and our unique circumstance.

At a territorial level, what steps should the Nunavut government take to mitigate the impact of climate change in Nunavut and the Arctic?

Nunavut is taking several steps to mitigate the impact of climate change in Nunavut and the Arctic.

For example, the government of Nunavut is developing and using information technology to centralize and increase access to climate change information, such as permafrost data and landscape hazards maps. This information is used to improve our infrastructure planning and determine which areas are being hit hardest by climate change.

In addition, the government of Nunavut is working through its recently established Climate Change Secretariat to increase adoption of energy conservation practices, reduce energy waste, reduce gas emissions and develop a concrete climate change mitigation strategy.

Nunavut and the government of Canada are also assessing the feasibility of electrification through hybrid power instead of diesel generation in Nunavut's communities. This could potentially reduce emissions while ensuring that Nunavut's communities [have] a reliable power source.

Can you personalize the impact of climate change by describing how the people of a specific town in Nunavut will be affected?

As a government, we have been working with partners to gather what we call "voices from the land." We value Inuit traditional knowledge, which is based on our close relationship with our land and environment. Inuit observations provide useful information at different points in time that contribute to our understanding of climate change.

Instead of providing just one example, I encourage your readers to view our website at http://www.climatechangenunavut.ca/en/voices-land, where they can navigate a plethora of Inuit voices from our land. It is really a compendium of interviews with elders, hunter and community members that are adding to scientific discussion through Inuit traditional knowledge.

What is your reaction to the UNIPCC's accepted calculation by Prof. Myers of a minimum of 200 million climate change refugees by 2050 and the world's readiness to cope with this humanitarian and environmental challenge.

This question has generated many arguments for and against. It isn't something I'd be in a position to address, as I am not an expert on human migration patterns or in humanitarian relief as a result of environmental challenges.

In Canada's north, essentially north of the 60th parallel, I would suggest northerners generally would prefer not migrate at any given time. There is a deep appreciation for our land and the northern way of life, and I'm sure there are similar views from those who have migrated due to natural disasters, such as flooding. This is why it is imperative that nations focus on adaption and mitigation and make real and concerted efforts to meet targets set out in the Paris Climate Change Accord. There is no one solution and there is no one body that can create change. It is going to take the global collective to ensure that we can accommodate the effects of climate change in every shape and form.

Interview

Alan Davis, Ph.D. is president and vice-chancellor at Kwantlen Polytechnic University (KPU) in Canada. He has worked for 40 years in higher education in Canada and the USA. He was born and raised in the

United Kingdom and attended University College London. He came to Canada in 1972 and received his Ph.D. in Chemistry from Simon Fraser University.

How can the university sector in Canada support innovation and projects on climate change mitigation?

It seems to me that universities are already central to the matter through research, education, advocacy, as places for debate and discussion, a place where students can organize and so on. They should also be exemplars of sustainability in all its forms on their campuses and in their practices. There are many networks and resources where best practice are shared and benchmarking can occur. KPU is doing its best within its limited resources to coordinate environmental sustainability initiatives, all of which tie directly or indirectly to climate change. We will need to have more dedicated leadership in order to move the dial any further.

What concerns you the most about climate change in your country?

There is still an extraordinarily poor understanding of how consumption and growth and "standards of living" impact climate change. So, we recycle and hug the trees, but then drive SUVs to go shopping.

Even though cars and trucks are becoming lighter and more efficient, they also expand in size, so overall the same energy is expended and we are not driving less. Public transit is a joke.

Also we don't really pay the full cost of our lifestyle, nor do we understand it. The average well-meaning Canadian family that recycles and so on and considers itself to be environmentally aware has no understanding that its lifestyle is still wildly unsustainable. The family does not understand the interconnectedness of consumer society and how it affects carbon. If you eat meat and/or drive a car, you are a serious part of the problem.

What concerns you the most about climate change globally?

If the developed world (especially the USA) doesn't take the lead, we cannot expect others to change. But the developed world is not taking the lead and there is no sign that it will. Thus it is already too late and things will get much worse before they get better.

Can you describe a specific sustainability project in your area that is contributing to a "greener world" and lower carbon footprint?

Not really. We are blessed in BC with an abundance of hydropower (which has its own limitations and impacts) and we have no concentration of heavy industry so we likely look good on paper regarding our carbon footprint.

However, almost nothing is being done about car and truck emissions (ask the people of Chilliwack after a few days of hot weather—that is where the smog blows to). There is no leadership with public transit. Vancouver has the worst traffic in the country with thousands of vehicles emitting carbon every day while going nowhere.

We talk a lot about clean and green technologies and we are the home of Greenpeace, but I see no over-arching strategy or goal with carbon. Plus, we still dig carbon up or cut it down and ship it to the beach to sell. Since the BC economy is so tied to the resource sectors, this is unlikely to change.

What steps could the Canadian government take to mitigate the growing threats from climate change?

1. Declare it to be the biggest issue facing the world.
2. Set a target that makes sense and that leads by example.
3. Develop incentives and policies to reach that target.
4. Organize and focus all government activities around this goal or none of this will happen.

What is your reaction to the UNIPCC's accepted calculation by Prof. Myers of a minimum 200 million climate change refugees by 2050 and the world's readiness to cope with this humanitarian and environmental challenge?

The IPCC is rarely wrong. I know people who work on the IPCC and they are thoughtful and careful scientists. They have no agenda except to seek the truth and advise accordingly.

Describe a particular area of climate change that you are concentrating upon at KPU with research or development projects.

We have proposed a partnership with industry and the city of Surrey to develop a clean and green technology accelerator. It was our top priority in our application to the Canadian and provincial governments for infrastructure funding. It was not funded, and no-one knows why: it met all the stated criteria. We continue to look for ways to leverage our land and resources to support this industry.

Interview

The Honorable Thomas Mulcair, former opposition leader, New Democratic Party, parliament of Canada, former environment minister, province of Quebec.

Comments

My primary work on climate change was in my capacity as Quebec minister of the environment for three years from 2003 to 2006. Quebec is Canada's largest province by area and second largest by population. We applied a rigorous plan to reduce greenhouse gases and in each of the three years that I was minister we succeeded in reducing GHG production in Quebec.

While a provincial minister I was given the unique opportunity to be the Canadian speaker at the plenary of the COP 9 in Milan, Italy, in December 2003. I was also instrumental in bringing together other subnational governments, such as Scotland and Wallonie, intent on playing a positive role in reducing GHGs during the COP 11 in Montreal in 2006. I authored Canada's most forward-looking legislation on sustainable development. The Sustainable Development Act was adopted unanimously by the National Assembly in 2006. One of its groundbreaking features was an amendment to the Quebec Charter of Rights that included the right to live in a clean environment. This provision has given rise to jurisprudence enabling the protection of ecosystems and biodiversity in the province.

In opposition in Ottawa since leaving Quebec provincial government, and as leader of the New Democratic Party for the last five years, I have promoted a strong vision of sustainable development that would include a concrete plan to meet our international obligations to reduce GHGs. My greatest concern is that Canada has consistently failed to act to reduce GHGs. After signing the Kyoto Protocol the then Liberal government went on to have one of the worst records in the world for GHG increases. No surprise then when a senior official from that government admitted that they had no plan and had signed to "galvanize public opinion." The Conservative government was even worse and made Canada the only country in the world to withdraw from the Kyoto Protocol. Back in power, the Liberals have just signed the Paris Accord while at the same time approving large energy projects that will make it impossible to reduce our GHG emissions. The Liberals have not produced the required economy-wide plan to meet Canada's Paris obligations and, contrary to their promises, they have only adopted the woefully inadequate plan of the previous Conservative government.

One hopeful initiative is the signing of a cap-and-trade agreement among several large sub-national governments, including Quebec and California. We have to replace all fossils fuels in electricity production without delay. Transportation remains a major source of GHGs in Canada, and electrification of public transit and moves to zero emission vehicles—with green renewable electricity sources—are part of the solution. One of the most vexing problems facing the world is that the climate change crisis doesn't yet feel like a crisis to many people. As a result, many governments continue to "fake it" and public pressure is not yet strong enough to get them to act. The challenge is to have enough people understand that we have to start acting now before a catastrophe makes it too late. At the same time, it is essential to diversify our economy by growing other industries. This will, of course, mean becoming leaders in renewable energy and energy saving technology, but will also include other economic sectors that have low environmental impacts but strong multiplier effects. We need to stop subsidizing fossil fuels and start supporting cleaner sectors.

Notes

1. IPCC, 2014: Summary for policymakers. In: Climate Change 2014: Impacts, Adaptation, and Vulnerability. Part A: Global and Sectoral Aspects. Contribution of Working Group II to the Fifth Assessment Report of the Intergovernmental Panel on Climate Change [Field, C.B., V.R. Barros, D.J. Dokken, K.J. Mach, M.D. Mastrandrea, T.E. Bilir, M. Chatterjee, K.L. Ebi, Y.O. Estrada, R.C. Genova, B. Girma,

E.S. Kissel, A.N. Levy, S. MacCracken, P.R. Mastrandrea, and L.L. White (eds.)]. Cambridge University Press, Cambridge, United Kingdom and New York, NY, USA, pp. 1–32.
2. Canada's Sixth National Report on Climate Change 2014 Actions to Meet Commitments under the United Nations Framework Convention on Climate Change 2014 Government of Canada.
3. Canada's Sixth National Report on Climate Change 2014 Actions to Meet Commitments under the United Nations Framework Convention on Climate Change 2014 Government of Canada.
4. Canada's Sixth National Report on Climate Change 2014 Actions to Meet Commitments under the United Nations Framework Convention on Climate Change 2014 Government of Canada.
5. Daniel Henstra and Jason Thistlethwaite "Climate Change, Floods, and Municipal Risk Sharing in Canada" IMFG Papers on Municipal Finance and Governance, No. 30, 2017.
6. Yusa, Anna et al. "Climate Change, Drought and Human Health in Canada." Ed. Jan C. Semenza. International Journal of Environmental Research and Public Health 12.7 (2015): 8359–8412. PMC. Web. 12 March 2017.
7. Daniel Henstra and Jason Thistlethwaite "Climate Change, Floods, and Municipal Risk Sharing in Canada" IMFG Papers on Municipal Finance and Governance, No. 30, 2017.
8. Natural Resources Canada, "Overview of Climate Change in Canada." http://www.nrcan.gc.ca/environment/resources/publications/impacts-adaptation/reports/assessments/2008/ch2/10321.
9. George McBean, "Telling the Weather Story: Can Canada Manage the Storms Ahead," Insurance Bureau of Canada, 2012.
10. Warren, F.J. and Lemmen, D.S., editors (2014): Canada in a Changing Climate: Sector Perspectives on Impacts and Adaptation; Government of Canada, Ottawa, ON, 286p.
11. Table 1 Key health concerns from climate change in Canada, Adapted from Séguin, J. (2008): Human health in a changing climate: a Canadian assessment of vulnerabilities and adaptive capacity; Health Canada, Ottawa, Ontario.
12. North America Chapter Intergovernmental Panel on Climate Change (IPCC), WGII 5th Assessment Report www.ipcc.ch.
13. "Upagiaqtavut, Setting the Course: Climate Change Impacts and Adaptation in Nunavut" Government of Nunavut.

14. James E. Overland and Muyin Wang "When will the summer Arctic be nearly sea ice free?" Geophysical Research Letters, Volume 40, Issue 10, 2013.
15. Peter Wadhams, Published at the Yale School of Forestry & Environmental Studies Yale University School of Environmental and Forestry Studies, September 16, 2016.
16. Climate Change Adaption Resource Guide, Government of Nunavut, Department of Environment, 2014.
17. Climate Action Tracker—Canada November, 2016. http://climateactiontracker.org/countries/canada.html.
18. Climate Tracker—Canada 2016. http://climateactiontracker.org.
19. Canada's Sixth National Report on Climate Change 2014 Actions to Meet Commitments under the United Nations Framework Convention on Climate Change 2014 Government of Canada.
20. Melillo, J.M., T.C. Richmond and G.W. Yohe, eds. 2014, Climate Change Impacts in the United States: The Third National Climate Assessment, U.S. Global Research Program, 841pp.
21. Walsh, J., and others. 2014: Ch. 2., Our Changing Climate. Climate Change Impacts the U.S.: The Third National Climate Assessment: J.M. Melillo, T.C. Richmond, and G.W. Yohe, eds. 2014, Climate Change Impacts in the United States: The Third National Climate Assessment, U.S. Global Research Program, 841pp.
22. Moser, S.C., M.A. Davidson, P. Kirshen, P. Mulvaney, J.F. Murley, J.E. Neumann, L. Petes, and D. Reed, 2014: Ch. 25: Coastal Zone Development and Ecosystems. Climate Change Impacts in the United States: The Third National Climate Assessment, J.M. Melillo, Terese (T.C.) Richmond, and G.W. Yohe, Eds., U.S. Global Change Research Program, 579–618. https://doi.org/10.7930/J0MS3QNW.
23. USGCRP (2009). *Global Climate Change Impacts in the United States*. Thomas R. Karl, Jerry M. Melillo, and Thomas C. Peterson (eds.). United States Global Change Research Program. Cambridge University Press, New York, NY, USA.
24. USGCRP (2014). Garfin, G., G. Franco, H. Blanco, A. Comrie, P. Gonzalez, T. Piechota, R. Smyth, and R. Waskom, 2014: Ch. 20: Southwest. Climate Change Impacts in the United States: The Third National Climate Assessment, J.M. Melillo, Terese (T.C.) Richmond, and G.W. Yohe, Eds., U.S. Global Change Research Program, 462–486.

25. USGCRP (2014). Walsh, J., D. Wuebbles, K. Hayhoe, J. Kossin, K. Kunkel, G. Stephens, P. Thorne, R. Vose, M. Wehner, J. Willis, D. Anderson, S. Doney, R. Feely, P. Hennon, V. Kharin, T. Knutson, F. Landerer, T. Lenton, J. Kennedy, and R. Somerville, 2014: Ch. 2: Our Changing Climate. Climate Change Impacts in the United States: The Third National Climate Assessment, J.M. Melillo, Terese (T.C.) Richmond, and G.W. Yohe, Eds., U.S. Global Change Research Program, 19–67.
26. USGCRP (2014). Garfin, G., G. Franco, H. Blanco, A. Comrie, P. Gonzalez, T. Piechota, R. Smyth, and R. Waskom, 2014: Ch. 20: Southwest. Climate Change Impacts in the United States: The Third National Climate Assessment, J.M. Melillo, Terese (T.C.) Richmond, and G.W. Yohe, Eds., U.S. Global Change Research Program, 462–486.
27. Ibid.
28. USGCRP (2014). Carter, L.M., J.W. Jones, L. Berry, V. Burkett, J.F. Murley, J. Obeysekera, P.J. Schramm, and D. Wear, 2014: Ch. 17: Southeast and the Caribbean. Climate Change Impacts in the United States: The Third National Climate Assessment, J.M. Melillo, Terese (T.C.) Richmond, and G.W. Yohe, Eds., U.S. Global Change Research Program, 396–417.
29. Ibid.
30. USGCRP (2014). Carter, L.M., J.W. Jones, L. Berry, V. Burkett, J.F. Murley, J. Obeysekera, P.J. Schramm, and D. Wear, 2014: Ch. 17: Southeast and the Caribbean. Climate Change Impacts in the United States: The Third National Climate Assessment, J.M. Melillo, Terese (T.C.) Richmond, and G.W. Yohe, Eds., U.S. Global Change Research Program, 396–417.
31. NOAA (2011). *Ocean and Coastal Resource Management in Your State: States and Territories Working With NOAA on Ocean and Coastal Management.* National Oceanic and Atmospheric Administration. Accessed 6/15/2011.
32. Munich Re. (2011, February). Topics Geo Natural catastrophes 2010: Analyses, assessments, positions. Retrieved May 19, 2011, from http://bit.ly/i5zbut.
33. Daniel Huber, Jay Gulledge, "Extreme Weather and Climate Change" Center for Climate and Energy Solutions, December 2011.

34. Karl, T.R., G.A. Meehl, T.C. Peterson, K.E. Kunkel, W.J. Gutowski, Jr., and D.R. Easterling, 2008: Executive Summary. Weather and Climate Extremes in a Changing Climate. Regions of Focus: North America, Hawaii, Caribbean, and US Pacific Islands. A Report by the U.S. Climate Change Science Program and the Subcommittee on Global Change Research, T.R. Karl, G.A. Meehl, C.D. Miller, S.J. Hassol, A.M. Waple, and W.L. Murray, Eds., 1–9.
35. Schwalm, C.R., C.A. Williams, K. Schaefer, D. Baldocchi, T.A. Black, A.H. Goldstein, B.E. Law, W.C. Oechel, K.T. Paw, and R.L. Scott, 2012: Reduction in carbon uptake during turn of the century drought in western North America. *Nature Geoscience*, **5**, 551–556. https://doi.org/10.1038/ngeo1529.
36. Karl, T.R., B.E. Gleason, M.J. Menne, J.R. McMahon, R.R. Heim, Jr., M.J. Brewer, K.E. Kunkel, D.S. Arndt, J.L. Privette, J.J. Bates, P.Y. Groisman, and D.R. Easterling, 2012: U.S. temperature and drought: Recent anomalies and trends. *Eos, Transactions, American Geophysical Union*, **93**, 473–474. https://doi.org/10.1029/2012EO470001.
37. USGCRP (2016). The Impacts of Climate Change on Human Health in the United States: A Scientific Assessment. Crimmins, A., J. Balbus, J.L. Gamble, C.B. Beard, J.E. Bell, D. Dodgen, R.J. Eisen, N. Fann, M. Hawkins, S.C. Herring, L. Jantarasami, D.M. Mills, S. Saha, M.C. Sarofim, J. Trtanj, and L. Ziska, Eds. U.S. Global Change Research Program, Washington, DC.
38. USGCRP (2014). Climate Change Impacts in the United States: The Third National Climate Assessment. Melillo, Jerry M., Terese (T.C.) Richmond, and Gary W. Yohe (eds.). United States Global Change Research Program.
39. USGCRP (2016). Impacts of Climate Change on Human Health in the United States: A Scientific Assessment. Crimmins, A., J. Balbus, J.L. Gamble, C.B. Beard, J.E. Bell, D. Dodgen, R.J. Eisen, N.Fann, M.D. Hawkins, S.C. Herring, L. Jantarasami, D.M. Mills, S. Saha, M.C. Sarofim, J. Trtanj, and L. Ziska, Eds. U.S. Global Change Research Program, Washington, DC. 312pp.
40. Center for Disease Control, CDC (2009).
41. USGCRP (2014). Climate Change Impacts in the United States: The Third National Climate Assessment. Melillo, Jerry M., Terese (T.C.) Richmond, and Gary W. Yohe (eds.). United States Global Change Research Program.

42. U.S. Census Bureau (2011). *The Next Four Decades—The Older Population in the United States: 2010.* U.S. Census Bureau. P2S–1138.
43. Environmental Protection Agency, "Climate Change and the Health of Children" 430-F-16-055 May, 2016.
44. Roger, V.L., and others, 2012: Heart Disease and Stroke Statistics—2012 Update: A Report From the American Heart Association. *Circulation*, 125, e2–e220.
45. Finkelstein, E.A., O.A. Khavjou, H. Thompson, J.G. Trogdon, L. Pan, B. Sherry, and W. Dietz, 2012: Obesity and severe obesity forecasts through 2030. *American Journal of Preventive Medicine*, **42**, 563–570.
46. Brault, M.W., 2012: Americans With Disabilities: 2010. 23pp., U.S. Census Bureau, Washington, DC.
47. Waidmann, T.A., and K. Liu, 2000: Disability trends among elderly persons and implications for the future. The Journals of Gerontology Series B: Psychological Sciences and Social Sciences, **55**, S298–S307.
48. USGCRP (2014). Hatfield, J., G. Takle, R. Grotjahn, P. Holden, R.C. Izaurralde, T. Mader, E. Marshall, and D. Liverman, 2014: Ch. 6: Agriculture. Climate Change Impacts in the United States: The Third National Climate Assessment, J.M. Melillo, Terese (T.C.) Richmond, and G.W. Yohe, Eds., U.S. Global Change Research Program, 150–174.
49. USDA (2016). Economic Research Service, undated. What is Agriculture's Share of the Overall US Economy?
50. USGCRP (2014). Hatfield, J., G. Takle, R. Grotjahn, P. Holden, R.C. Izaurralde, T. Mader, E. Marshall, and D. Liverman, 2014: Ch. 6: Agriculture. Climate Change Impacts in the United States: The Third National Climate Assessment, J.M. Melillo, Terese (T.C.) Richmond, and G.W. Yohe, Eds., U.S. Global Change Research Program, 150–174.
51. Mathew E. Hauer, Jason M. Evans, Deepak R. Mishra, "Millions projected to be at risk from sea-level rise in the continental United States" in Nature Climate Change, 6, 691–695, (2016).
52. *Climate Change Adaptation in New York City: Building a Risk Management Response,* Michael R. Bloomberg, Jeffrey Sachs, Gillian M. Small, 24 May 2010.
53. https://phys.org/news/2015-02-nasa-science-york-city-climate.html#jCp.
54. Climate Tracker—United States, 2016. http://climateactiontracker.org.

3

South America: Brazil, Ecuador, Argentina

"We have no other spare or replacement planet. We have only this one and we have to take action."
Berta Caceras

Introduction

South America is the fourth largest continent in the world with a land mass of approximately 6,878,000 square miles (17,814,000 km^2). The highest elevation is Mount Aconcagua in Argentina at 22,831 feet (6959 m). The coastline totals 15,800 miles. Climate change impact on South America is significant and expected to increase in intensity across all regions. Noticeable trends include elevated heat levels, drought, erratic rainfall patterns, flooding, rising sea levels, coastal flooding, saltwater intrusion, mild to severe damage to agricultural lands depending upon the regions, negative health impacts and sustained infrastructure repair costs.

The IPCC has noted the following outcomes that will affect the region in general. "Significant trends in precipitation and temperature have been observed in Central America (CA) and South America (SA) with high confidence; extreme events have severely affected the region, with medium confidence; increasing trends in annual rainfall in southeastern South America; warming has been detected throughout CA and SA; increases in temperature extremes have been identified in CA and most of tropical and subtropical SA with medium confidence; more frequent extreme rainfall in southeastern SA has favored the occurrence of landslides and flash floods with medium confidence; warming varies from +1.6 °C to +4.0 °C in CA, and +1.7 °C to +6.7 °C in SA with medium confidence; by 2100 projections show an increase in dry spells in tropical SA east of the Andes, and in warm days and nights in most of SA with medium confidence."[1]

There is a direct and scientifically verified correlation between greenhouse gas (GHG) emissions and climate change. The enormity of the task of reducing CO_2 emissions globally was noted in a recent economic report analyzing climate change impacts in Central and South America. "It is thought that a determined effort to reduce the emissions level from slightly less than 7 to 2 tons per capita by 2050 will have to be made in order to stabilize climate conditions in a world where most economies are heavily reliant on fossil fuels. The economic and social challenge of devising ways to deal with the economic, social and environmental losses and costs associated with climate change, while at the same time mitigating the effects of greenhouse gas emissions, will shape the development style of the twenty-first century."[2]

One of the enormous challenges for governments across the region is to combat poverty, uplift socio-economic development while simultaneously directing billions of dollars to climate change mitigation and adaption strategies. This task will be rendered more difficult as climate change effects increase in severity over the coming decades according to climate science projections.

Brazil

Brazil has a population of 210 million and is the largest country in South America with a total area of 8,515,767.049 km² (3,287,956 square miles) including 55,455 km² (21,411 square miles) of water. The land area of

Brazil comprises 47% of South America. It ranks ninth in global economic standing with a GDP of USD 2 trillion in 2016 and 75th in the Human Development Index (HDI). CO_2 emissions in 2016 were 1.35% of the global total. It is a member of the BRIC trade bloc that includes Brazil, Russia, India and China.

In recent years, Brazil has faced intense international criticism from environmentalists over the rapid rate of deforestation in the ecologically diverse and rich Amazon rainforest, which is the largest in the world. The Amazon rainforest covers an area approximately the size of the continental USA. Only 3% of the Earth's land is covered by rainforest. There are an estimated 390 billion different trees grouped into 16,000 species in the Amazon rainforest that is dated at 55 million years old. Brazil contains 64% of the rainforest with Ecuador, Venezuela, Surinam, Peru, Colombia, Bolivia, Guyana and French Guiana occupying the remaining portions. Over the past 40 years approximately 20% of the Brazilian rainforest has been cut down, which has had a devastating effect upon the environment in four notable ways. First, the rainforest mitigates GHG emissions thus extensive deforestation is raising CO_2 emissions resulting in global warming. Second, water cycles are being severely disrupted. Third, soil erosion is accelerated along with a decline in soil moisture due to the disappearance of thousands of trees and millions of plants. Indeed moisture levels have fallen significantly in areas several miles away from the deforested areas thus damaging ecologies and contributing to drought. Fourth, since about 80% of global species are found in tropical rainforests, deforestation is destructive to habitat and thousands of species. The Brazilian government has taken solid steps in the past decade to reduce deforestation and this is having a slow but restorative effect upon ecological preservation and the environment.

Flooding and Drought

Although drought and flooding have been constants in Brazilian recorded history the rate of these climate phenomenon appear to be increasing in length and intensity. A climate change risk report by global insurer Lloyds of London noted the following sea level rise and coastal flooding scenarios for the major port cities of Recife and Rio de Janeiro. "It is for this reason that variations to the order of 15 cm for Recife and 30 cm for Guanabara Bay in Rio de Janeiro have been estimated for 2030. Due to the extent of the Brazilian coastline, the factors that intervene in this narrow coastal strip (physical, Alteration in coastal bathymetric profile climatic, oceanographic, biological and man-made) and affect the magnitude of the impacts from rising sea levels are diverse. The combination of these factors promotes several risk scenarios."[3] Such a scenario will have a profound impact on infrastructure, health, water delivery, agriculture and commerce.

> Under a high emissions scenario, and without large investments in adaption, an annual average of 618,000 people are projected to be affected by flooding due to sea level rise between 2070 and 2100. If emissions decrease rapidly and there is a major scale up in protection (i.e., continued construction/raising of dikes) the annual affected population could be limited to about 3200 people. Adaption alone will not offer sufficient protection, as sea level rise is a long-term process, with high emissions scenarios bringing increasing impacts well beyond the end of the century.[4]

In one two-year period, 2013–2015, Brazil experienced the worst drought in a century. There were 1485 municipalities and townships that declared states of emergency out of 5561 in the entire country. São Paulo, the largest city in the country with a population of 12 million, faced severe water shortages and disruption of vital services. The main water system, the Cantareira reservoir, is only built to support the water requirements of 5 million people, yet reached record low levels at 17% of capacity, thus creating a water scarcity emergency. Such situations of water insecurity are symptomatic of the erratic weather patterns, drought and flood scenarios that can be heightened by climate regular water supply for several months, due to strict water rationing imposed by the state government. The drought also had an impact on industrial output, agriculture and public health. Indeed, 2014 was one of the driest recorded years in São Paulo's history.

Drought is a recurring natural phenomenon in Brazil. The incidence of drought appears to be increasing and the long-term projections do not offer an improvement in this scenario. According to research from the Brazilian Atlas of Natural Disasters, "Between 1991 and 2010, there were close to 17,000 drought events recorded in 2944 municipalities in the country, making it the top disaster by type, with over 50% of total disaster events reported. Of a total of 96 million affected persons in these 20 years, 48 million (50%) were affected by drought (flash floods and other floods made up to 40%); and over a total of 2475 registered deaths, roughly 10% (257) were drought related."[5]

In 2015, Brazil struggled to supply enough water to its 200 million people against the backdrop of the worst drought in nearly a century. São Paulo's 20 million citizens face having their tap water cut off five days a week in a bid to conserve dwindling resources. Some 17% of Brazilian towns have declared a state of emergency. In the center and southeast of the country, electricity supplies are threatened as water levels in hydropower generation reservoirs drop to 18%. "According to the Center for Weather Forecasting and Climate Research (CPTEC/INPE), out of the 10 hottest temperatures registered worldwide on December 31, 2013, 9 occurred in Brazil."[6]

The northeastern region is semi-arid and particularly vulnerable to heat waves and drought. A predictive analysis for the region for the period 2025–2050 offered a sober assessment of expected outcomes. "The region is of particular relevance to the study of climate change impacts given its large human population (28% of Brazil's population) and high levels of impoverishment, having an extensive semi-dry area which will be severely impacted by growing temperatures."[7]

Extreme Weather

Extreme weather events are not new in Brazilian history yet are expected to increase in both frequency and intensity. In southern Brazil, heavy rainfall struck Santa Catarina State in November 2008, which caused devastating flooding and mudslides that affected 1.5 million people and rendered nearly 70,000 people homeless. The majority of the population lives in the south and southeast of the country and major cities such as Rio de Janeiro. In 2014, during the extreme drought, 17 of the country's

18 biggest reservoirs were at their lowest levels in years and the southeast was impacted the most negatively. The important Paraiba reservoirs in Rio de Janeiro were dangerously low at 1% capacity.

The available climate science data projections indicate that extreme weather events in Brazil will be particularly destructive in the southeast regions. The National Space Research Institute (INPE) and the Natural Disaster Surveillance and Early Warning Center (CEMADEN), which are associated with the Brazilian Instituto Nacional de Ciencia e Tecnologia para Mudancas Climaticas (INCT-MC), are projecting heavy rains and extreme droughts to become more frequent and intense in certain regions of Brazil, especially the south and southeast. "Climate scenarios produced for South America during the current century point to regional variations in climate change and its impacts. The projections indicate that parts of the north of the American continent will have insufficient rainfall while parts of the southeast will experience much heavier rainfall."[8] With regard to the drought in southeast Brazil, INCT-MC's researchers evaluated the specific features of extreme periods in the region and their correlations with sea surface temperature (SST) in the South Atlantic, as well as the related synoptic weather patterns, comprising observations of large-scale phenomena such as depressions, cyclones and anticyclones.[9]

Projections of changes in extra-tropical cyclones deriving from global warming scenarios—based on a range of assumptions for the trajectory of GHG emissions—suggest that cyclone formation areas near Brazil's southern coast will move southward between 2071 and 2085, affecting atmosphere blocking patterns over the subtropics and potentially causing droughts in several agricultural zones of Brazil.[10] Brazil is ranked among the 15 countries most vulnerable to sea level rise. This trend will be exacerbated by extreme weather events related to flooding and coastal erosion.

Health, Human Development and Food Security

Brazil's economic and social progress between 1990 and 2015 has been notable. Moreover, between 2003 and 2014 approximately 30 million citizens were lifted out of poverty. Yet poverty and inequality remain difficult problems. With regards to climate change impacts it is normally the impoverished, disabled, elderly, health challenged and children who are the most vulnerable. In 2017, 21% of the population lived below the poverty line defined as people living on less than USD 1 per day.

"Socioeconomic conditions have improved however, there is still a high and persistent level of poverty in most countries, resulting in high vulnerability and increasing risk to climate variability and change (high confidence). Poverty levels in most countries remain high (45% for CA and 30% for SA for year 2010) in spite of the sustained economic growth observed in the last decade."[11] In Brazil there are many health complications caused by climate change increasing the temperature in a country already noted for extremes of heat and episodic drought. Moreover, a myriad of diseases are accelerated in a tropical climate that is experiencing steadily warming temperatures. "Under a high emissions scenario heat-related deaths in the elderly (65+ years) are projected to increase to about 72 deaths per 100,000 by 2080 compared to the estimated baseline of about 1 death per 100,000 annually between 1961 and 1990 and by 2070, over 168 million people are projected to be at risk of malaria assuming a high emissions scenario. If emissions decrease rapidly, projections indicate this number could be limited to about 126 million."[12]

Despite the serious and, at times, fatal health consequences of climate change, the imperative targets to reduce GHGs and global warming in Brazil are not achievable. As noted in a recent report, "According to our analysis, Brazil's emissions reduction targets are at the least ambitious end of a fair contribution to global mitigation, and are not consistent with meeting the Paris Agreement's long-term temperature goal unless other countries make much deeper reductions and comparably greater effort. Our analysis shows that due to increasing energy demand and an implementation lag that affects climate policy in Brazil, emissions in most sectors are expected to continue rising until at least 2030."[13]

Brazil is considered relatively stable in terms of food security. Threats to food systems and agriculture will primarily affect the southeast where droughts and weather extremes are projected to increase. Inconsistent water security will have an impact upon agriculture in the northeast where the lowest farm incomes are found. There are a high number of subsistence farmers who are particularly vulnerable to shifting weather patterns. The southeast is a more developed region in terms of agriculture and has higher levels of technological implementation. Common crops include wheat, peas, beans, cotton and sugarcane. In the south, the main crops are maize, beans and soybean. It is a highly developed region.

Climate Change Refugees

As a highly developed industrial economy, Brazil has the adaptive capacity and resources to address internal migration from environmental crises. The country is not expected to generate external climate refugee flows. However, internal displacement due to drought, particularly in the southeast, flooding along coastal areas and heavy rainfall in the northeast and parts of the southeast will likely lead to increased internal migration. Demands for proactive government responses and emergency assistance will escalate in correlation with the severity of these internal migration flows.

City Profile: Rio de Janeiro

Rio de Janeiro is a leading city in Brazil with a geographical area of 1200 km² and 2017 population of 6.5 million. In 2012, it was designated a World Heritage Site by the United Nations Educational, Scientific and Cultural Organization (UNESCO). A large slum population who live in the city's 1000 favelas is estimated at 1.5 million people or 24% of the city's population. The favela inhabitants represent a highly impoverished and often undernourished population who are vulnerable to climate change and largely ill prepared to adapt to the associated disruptions to health, housing, employment and infrastructure.

The main climate change induced health threats to the city population include extreme heat, air pollution, respiratory problems, cardiovascular diseases, gastroenteritis and other infectious diseases such as dengue fever and leptospirosis.[14] Vulnerable groups include children, the elderly, the mobility challenged, the mentally ill and people with immune systems disorders. Coastal flooding, shoreline erosion and saltwater intrusion are additional threats that face the city. "The projections show that, by 2080, temperatures in Rio will increase by 3.4 degrees, the same increase predicted for the city of Cubatão in the state of São Paulo. The sea level is set to rise by between 37 cm and 82 cm over the next 65 years. Of the 92 municipalities in Rio de Janeiro State, the city of Rio de Janeiro is the most vulnerable to the climatic changes predicted over the course of the next three decades. On a scale from 0 to 1, the city of Rio is rated as 1, due to its vulnerability in the area of health and environment."[15]

Innovation and Adaption Strategies

Brazil signed the Paris Agreement in 2016. It is also committed by international treaties and domestic policy to reduce GHGs substantially. Government targets include, relative to 2005, an estimated reduction in terms of emissions intensity in 2025 by 70% and in 2030 by 79%. This represents a reduction of 48% in emissions by 2030. Brazil has voluntarily committed to reducing GHG emissions to between 36.1 and 38.9% by 2020.

The government has instituted a number of progressive initiatives to mitigate climate change. These steps include:

1. 2010 Brazil National Fund on Climate Change;
2. 2009 Brazil Policy on Climate Change;
3. 2002 Brazil ratified the Kyoto Protocol;
4. 1992 Brazil signed the UNFCCC.

The government has also embarked upon a consistent path toward reducing deforestation levels. This is important in reducing GHGs, global warming, the ecological destruction of thousands of species and protecting moisture levels. Indeed, drought is a recurring climate phenomenon facing Brazil and is expected to be exacerbated by climate change. In 2008, Norway pledged USD 1 billion in support to Brazil if the government effectively reduced deforestation and CO_2 emission levels. This program is part of a broader and innovative initiative called Reducing Emissions from Deforestation and Forest Degradation (REDD). In 2015, the government approved the Environmental Monitoring Program of the Brazilian Biomes. This new program is designed to monitor deforestation and evaluate a range of related environmental concerns such as land use and the protection of plants and vegetation.

Ecuador

Ecuador is situated in the northwest region of South America and has a population of 16.5 million (2017). It has a land mass of 283,561 km^2, a water area of 6720 km^2 and 2237 km coastline. Its GDP is $99 billion and has an HDI ranking of 88th in the world. As a Pacific Ocean nation, Ecuador is vulnerable to sea level rise, coastal flooding and saltwater intrusion. The country is also subject to floods, episodic drought and earthquakes. Approximately 30% of the land is dedicated to agriculture, which poses development challenges in the face of increasing climate change effects. Ecuador contributes approximately 0.1% of the world's GHG emissions but the effects of global warming are evident throughout the country, especially around sources of water. Between 1996 and 2008, for example, Ecuador's glacier coverage was reduced by 28%.

Flooding and Drought

From December 1997 through April 1998, heavy rains and rising sea levels caused extensive flooding along Ecuador's Pacific coast that forced the evacuation of 30,000 citizens, caused 286 fatalities and the loss of USD 1.5 billion in agriculture production. Damage to infrastructure was estimated at USD 830 million. Moreover, the economic loss from this weather crisis amounted to 15% of the country's GDP. Ecuador is extremely vulnerable to flooding. The four provinces facing the most serious risk are Los Ríos, Esmeraldas, El Oro and Guayas. However, population in these provinces is increasing, which places hundreds of thousands of citizens in heightened danger from the health, economic and infrastructure repercussions of extensive flooding. The combined impacts of both drought and flooding are constant human and environmental threats in the Daule region. This is of particular concern to rice farmers whose crops are sensitive to both of these weather impacts. Every year hundreds of acres of valuable rice lands are lost. In 2012, severe rainfall and consequent flooding caused disruption to 100,000 people and severely damaged rice crops and growing fields in Daule.

In a recent report, the consequences of several climate change scenarios facing Ecuador were evaluated with the following observations: "A temperature increase of 2 °C and a precipitation decrease of 15% would lead to agricultural shortages; and second, a temperature increase of 1 °C and an increase in precipitation of 20% would severely affect fisheries and agriculture production as a result of increased flood risk."[16] Drought is also a recurring threat to Ecuador. As temperatures are expected to rise slowly and steadily over the course of the twenty-first century, drought impact will intensify. Since 2005, drought has cost the economy an estimated USD 4 billion.

Extreme Weather

Extreme weather events that are expected to increase in force in the country include, drought, heavy rainfall and flooding. The El Niño of 1997–1998 resulted in the highest recorded precipitation in Ecuadorian history, which was more than double that of a normal rainy season. In sum, El Niño destroyed 843,873 ha of crop lands in Ecuador, and Guayas province endured damage to 26% of the cultivated land. Many climate scientists anticipate that the effects of El Niño will intensify in the twenty-first century. According to a 2016 UN report, in Ecuador hundreds of floods and landslides have devastated communities, the El Niño phenomenon has also wreaked havoc on global rainfall patterns, leading to disastrous flooding and prolonged droughts.[17]

Health, Human Development and Food Security

Flooding leads to many serious health problems. It can trigger highly elevated mosquito levels, which increases the risk of malaria. Other health crises precipitated by flooding include the spread of dengue fever and chikungunya fever. With high rates of poverty in regional areas and rural

Ecuador, climate change induced drought and flooding will have negative repercussions upon the poor, marginalized, the health challenged and vulnerable groups such as subsistence farmers. Common agricultural crops include corn, soybeans, potatoes and vegetables, which are typically grown in higher elevations, and other crops such as bananas, sugarcane, coffee and rice, which are cultivated in the lowland and coastal areas. Approximately 18% of the population is engaged in agriculture. In 2016, 24% of the population was living below the poverty line. Subsistence farmers are the group most vulnerable to income decline and job losses due to the environmentally induced damage to land and crops.

There are serious and quantifiable threats to food security and agricultural income levels due to climate change. As one report noted, "Climate-related disasters affect key sectors such as agriculture, water resources, fisheries, infrastructure and tourism, and especially affect rural areas with large populations of indigenous and Afro-Ecuadorian people. Climate variability, including more frequent and intense El Niño and La Niña phenomena, combined with pockets of food insecurity and poverty, have led Ecuador to prioritize sound planning and replicable implementation models to address these threats."[18]

"For agricultural workers, occupational exposure to hot environments is another major climate change-related problem. Indeed, agricultural workers are one of the working groups at highest risk of heat stress, followed by construction and manufacturing workers, miners, armed forces personnel and fire-fighters [4]. In the Ecuadorian coast, agricultural workers usually perform their activities under poor working conditions, at temperatures between 25 and 35 °C degrees and humidity of 60–80%, increasing the risk of heat stress."[19] Generally, rising temperatures in many regions of the country, episodic heat waves that are expected to increase in intensity and elevated UV levels are growing health and occupational concerns for millions of citizens.

In Quito, Ecuador, between 2006 and 2010, skin cancer showed an increased incidence and prevalence, placing this type of cancer as the second most common cancer among men (after prostate cancer) and women (after breast cancer), with an increasing trend over the last ten years.[20] "In the city of Cuenca, cases often occur in agricultural workers.

Studies in workers with outdoors occupations and exposure to UV radiation have reported an increased risk of skin cancer.[21,22] The high number of agricultural workers in Ecuador combined with significant poverty and harsh exposure to heat are significant factors that endanger health protection. Two of the most vulnerable groups are low income and subsistence farmers and farm workers.

Climate Change Refugees

As a developing industrial and agricultural economy with positive indicators, Ecuador will have the capacity to respond sufficiently to most climate refugee scenarios. It is not expected to generate external climate refugee flows. Internal migration from environmental crises will be caused by coastal flooding, drought episodes and extreme weather events such as intense rainfall. The National Strategy for Climate Change and other government initiatives are designed to provide support and information to citizens affected by internal environmental displacement.

Innovation and Adaption Strategies

Ecuador signed the Paris Agreement in 2016. It has signed but not ratified the following environmental agreements: Antarctic-Environmental Protocol, Antarctic Treaty, Bio-diversity, Climate Change, Climate Change–Kyoto Protocol, Desertification, Endangered Species, Hazardous Wastes, Ozone Layer Protection, Ship Pollution, Tropical Timber 83, Tropical Timber 94 and Wetlands.

The country is implementing the National Strategy for Climate Change, as per Decree 1815. This decree establishes the preparation and launch of the National Strategy for Climate Change for 2012–2025.

The Ecuadorian government in concert with regional authorities has inaugurated a major new project with the UN World Food Programme (WFP), and the UNFCCC Adaption Fund that is designed to build community and ecosystem resilience, reduce climate vulnerability and enhance food security.

Four provinces, Pichincha, Azuay, Loja and El Oro have been selected, containing two important watersheds. The prioritized area covers 12 cantons, 50 parishes, approximately 120 communities and 15,000 families. It encompasses varying ecological systems, cultural traditions, ethnic composition and dependency on natural resources. The project began in 2011 and operated until 2016 with the aim of collecting better climate information and identifying community priorities for climate change adaption and to build environmental sustainability. It is organized around two main goals:

1. to develop community-level awareness and knowledge on climate change and food insecurity related risks;
2. and to increase adaptive capacity for climate change.[23]

The National Plan for Good Living (2013–2017) is designed maximize the sustainable management of resources and bio-diversity and support strategies for the mitigation of climate change as well as effective adaption policies.

In 2013 the Ministerial Accord No. 33 on REDD+ established a regulatory framework for the implementation of the REDD+ mechanism in Ecuador.

The National Strategy on Climate Change (Executive Decree No. 1815 and Ministerial Accord No. 095, 17 July 2009) sets forth the framework for climate change adaption and mitigation as a formal government policy. Important elements of the strategy include: the protection of vulnerable groups and ecosystems; measures to enhance food security; measures to protect health; measures to protect water sustainability; enhanced disaster management approaches; improved responses and adaption of communities large and small to the effects of climate change.

Argentina

Argentina is situated on the west coast of South America with a land area of 2.8 million km² and a coastline of 4989 km. The 2017 GDP is USD 580 billion and the country ranks 40th in the HDI. Approximately 12% of the population of 44 million live in poverty, which the government defines as USD 4 per day. The CO_2 emissions are 0.53% of the global total.

Approximately 47% of Argentina's land is utilized for agriculture, which employs 5% of the labor force. The high agricultural land ratio renders a major portion of the economy and economic livelihood extremely vulnerable to the effects of climate change upon food and land systems. As a leading global food producer and agricultural nation it is vulnerable to the negative climate change effects of drought, extreme heat, sea level rise, flooding and saltwater intrusion upon land and resources.

Climate scenario projections to the period 2080/2090 have noted the following concerns. "(a) increases in temperature—it is probable that the temperature will increase by 1 °C in the whole country by 2020/2040, but particularly in the North, where agriculture is concentrated, leading to greater evaporation and subsequently to increased aridity and desertification. (b) changes in precipitation—precipitation increases are expected in the central part of Argentina, mainly during the summer and fall season, while a reduction in precipitation levels is projected for the Andes, Northeastern part of Patagonia and Comahue. Extreme precipitation events are expected to be more frequent. (c) rising sea level—it is probable that the average sea level will rise leading to coastal erosion, particularly in the Southern part of Patagonia."[24]

Moreover, rapidly rising temperatures are having an impact upon the glaciers of Argentina. The Upsala Glacier, one of the largest glaciers in the world, has been losing ice mass at a rate of 200 m a year and 8 km over the last 25 years. Similar to the rapid ice melt ratio in the Arctic, it is conceivable that by 2050 this monumental glacier will have largely disappeared.

Flooding and Drought

Climatic change is a real concern for Argentina because of its economic dependence on farming. Since the beginning of the twentieth century, natural events such as floods or droughts affecting farming triggered palliative measures involving tax reductions, low-interest-rate loans, and the like.

Climate change has a negative impact upon agricultural output and actually increases the propensity for natural disasters. One report has noted that without any substantial mitigation efforts or international aid the combined costs of infrastructure damage and lost agriculture revenue will require more than 10% of Argentina's GDP by 2030, and by the end of the century it will render the nation insolvent.[25] As early as 2007, in its second report to the UNFCCC, the government stated that the nation is vulnerable to climate change induced floods and landslides triggered by increased rainfall and melting glaciers.

Extreme Weather

Nearly 40 cm of rain fell within two hours in Buenos Aires and La Plata, Argentina, in April 2013—the heaviest rainfall recorded in the country in more than a century. The event claimed 57 lives and caused infrastructure damage and economic losses estimated at USD 1.3 billion. Buenos Aires had record rainfall levels in 2013 and Iguazu had record breaking flood levels. Over the past 40 years rainfall has increased 10% in the northeast and up to 40% in La Pampa Province. The addition of extreme rainfall has a negative impact on flood levels and hence agricultural lands, and also contributes to health problems, infrastructure damage and mudslides.

"Between 1999 and 2008 floods and storms have had the highest human and economic impact in Argentina, with losses averaging 0.23% of GDP—678,040 people have been affected by floods (9 events) with the cost of the damages reaching US$ 2.1 billion and 5000 people have been affected by storms (1 event) with damages reaching US$15 million. The damages of a drought event in June 2003 reached US$120 million."[26] "From December 13 to 31, 2013, a strong heat wave occurred over central Argentina with maximum temperatures over 40 °C and minima over 28 °C. This heat wave was the longest and the most intense, considering the accumulation of degrees over thresholds, ever registered in the region. Record values of minimum temperature were verified in a station close to Buenos Aires with an estimated 100-year return period."[27]

Health, Human Development and Food Security

As mentioned earlier, the impact on agriculture is the most common natural risk Argentina suffers, and a national strategy is needed to overcome or at least diminish its effect. This section of the chapter focuses mainly on agricultural risks generated by climate change. Argentina is a country where the agricultural sector plays an important role in the economy, with a substantial segment of the population employed in this vital sector. In Argentina, several institutions—including the Agricultural Risk Bureau (Oficina de Riesgo Agropecuario, ORA) and the National Institute of Agricultural Technology (Instituto Nacional de Tecnología Agropecuaria, INTA), the Sevicio Meteorologico Nacional (SMN), and the Instituto Nacional del Agua (INA)—develop and guide policy related to the agriculture sector and its sustainability. Some of the climate change impacts upon agriculture and food security have been well noted. According to a leading study, "yield losses in the north are projected, and the northern and northeastern parts of the country could suffer a desertification process due to increased evapotranspiration from increased temperatures and lack of precipitation, rendering the regions unsuitable for agriculture."[28]

Climate Change Refugees

Argentina is a highly developed industrial economy. It is not projected to generate external refugee flows. It will however have to contend with isolated regional challenges of human displacement due to extreme heat episodes that intensify drought and flooding problems in vulnerable coastal areas. The high economic standing of the country will permit more successful mitigation and adaption strategies compared to its regional neighbors that are at a lower economic and infrastructural development level.

Innovation and Adaption Strategies

In 2016 Argentina ratified the Paris Agreement.

In October 2015, Law 27,191 was introduced. It amended Law 26,190 (2007 Regimen for the National Promotion for the Production and Use of Renewable Sources of Electric Energy) and introduced new provisions to promote the development of renewable energy sources. Law 27,191 established stringent targets on the proportion of Argentina's electricity supply to come from renewable sources: 8% by December 31, 2017; 12% by December 31, 2019; 16% by December 31, 2021; 18% by December 31, 2023; and 20% by December 31 2025.

On October 1, 2015, Argentina submitted its Intended Nationally Determined Contribution (INDC), including a goal to reduce GHG emissions including land use, land use change and forestry (LULUCF) by 15% below its business-as-usual (BAU) scenario by 2030 (equivalent to 60% above 2010 levels or 128% above 1990 levels excluding LULUCF).

In 2009, Act No. 26,509 created the National System of Prevention and Mitigation of Farming Emergencies and Disasters.

In 2009 the Agricultural Emergency Act (Ley de Emergencia Agropecuaria, Act No. 26,509, August 2009) created the National System for Prevention and Mitigation of Agricultural Emergencies and Disasters. The objective of the system is to prevent and mitigate the damage caused by climate factors that significantly affect agricultural production and rural communities, directly or indirectly.

In 2009, Argentina became a member of the UNREDD program. It has a budget of USD 3.59 million beginning in 2014. The "Roadmap for Readiness" has four objectives: (1) to draft a REDD+ National Strategy, (2) to establish a National Forest Reference Emissions Level, (3) to strengthen the National Forest Monitoring System, and (4) to develop a Safeguard Informational System.

The 2007 Regimen for the National Promotion for the Production and Use of Renewable Sources of Electric Energy (Law 26.190) states that the production of electricity from renewable energy sources is in the national interest.

The Policy Decree 140/2007 is a presidential decree declaring "rational and efficient" energy use as a national priority. The decree was designed to reduce energy consumption and promote the use of renewable energy in the public sector, private sector and in residential dwellings.

The Promotion of Hydrogen Energy (Law 26.123 [2006]) states that the technological development, production and use of hydrogen fuel, as well as other alternative energy sources, is in the national interest.

The Regimen of Regulation and Promotion of the Production and Sustainable Use of Bio-fuel (Law 26.093 [2006]) provides a regulatory framework for the production and promotion of bio-fuel. Its goal is to have all gasoline produced and consumed in Argentina composed of no less than 5% bio-fuel.

Policy National Decree 1070/05 (2005) established the Argentine Carbon Fund. Its goal is to promote projects in keeping with the intent of The Clean Development Mechanism noted in Article 12 of the Kyoto Protocol.

Law 25019 (1998) states that wind and solar power are in the national interest. The law was designed to promote renewable energy. The Renewable Energy Fund was established to promote production.

Interview

Isabel Alvarez is an environmental engineer specializing in water resource management, supply and sanitation. She has worked in four different provinces in Ecuador and in Canada, where she managed water and sanitation programs and projects that served the most vulnerable communities. She is the author of the published article "Analysis of the Viability of Ethanol Production in Brazil: Economical, Social & Environmental Implications." She completed a master's degree in Integrated Water Resource Management at McGill University.

Can you describe your work related to climate change.

As global temperatures increase and rainfall patterns shift due to the effects of climate change, severe weather events such as flooding and drought are becoming increasingly frequent. Droughts can have disastrous consequences since water scarcity will undermine agricultural activities and people's health in general. My work in the past three years has focused on providing clean drinking water and adequate sanitation to people living in poor and marginalized areas in Ecuador. Accessing these basic services is paramount to fight water, sanitation and hygiene related diseases that kill nearly 1 million people each year worldwide.

What concerns you the most about climate change in your country?

One of the most concerning effects of climate change in Ecuador is the El Niño phenomenon, which not only affects Ecuador but also other

South American countries. In 1997, the Ecuadorian government declared a state of emergency due to heavy flooding in many towns and cities of the coastal area of the country. Many people lost their homes and 286 people died as result of infectious diseases triggered by the flooding. According to the Climate Prediction Center, historic data shows that the two most recent El Niño events have been the strongest. Moreover, glacier melting in Ecuador is one of the most noticeable effects of climate change. Between 1996 and 2008, Ecuador's glacier coverage went down by 28%. The water that comes from Cotopaxi's glaciers, one of the most active volcanoes in the world, is the main source of water for Ecuador's capital city, Quito. If this trend continues, millions of people will suffer from water scarcity.

What concerns you the most about climate change globally?

There are two main concerns: global food supply and climate change refugees. The change in rainfall patterns and the evaporation of aquifers could cause a shift in agricultural practices from rain-fed to irrigated agriculture. These new practices would not only further exacerbate water scarcity but also have a negative impact on global food supply. Furthermore, severe weather events will displace populations in massive numbers in areas with limited water resources thus creating more demand.

Can you describe a specific sustainability project in your area or country that is contributing to a "greener world" and a lower carbon footprint?

The In Terris foundation based in Guayaquil, Ecuador, is promoting sustainable sanitation solutions through the use of an EcoSan toilet that does not require water to function and that transforms waste into compost to be used on agricultural land. This toilet has been introduced in some villages of Ecuador and it has been well received by many. Nevertheless, some people have misgivings in regards to the safety of handling human waste and using it as fertilizer for food crops. Overall, the EcoSan toilet is a viable alternative to conventional toilets since it eliminates the need for water, while significantly reducing water contamination, and it creates added value in the form of fertilizer.

What steps could your government take to mitigate the growing threats from climate change?

Ecuador was the first country in the world to recognize the rights of nature in its constitution, which is a significant step forward to fight climate change. However, in practice the country still relies on fossil fuel for its source of energy, which measured 89% in 2011. Moreover, deforestation occurs at a rate of 66,000 ha per year, in part due do illegal logging. Ecuadorian authorities have tried to combat this problem by means of programs such as Socio Bosque, which offers cash incentives to locals to preserve forest. I believe that Ecuador is on the right path in terms of creating policies to protect the environment, but the implementation of these policies is lacking.

What is your reaction to the UNIPCC's accepted calculation by Prof. Myers of a minimum of 200 million climate change refugees by 2050 and the world's readiness to cope with this humanitarian and environmental challenge.

Even though Prof. Myers prediction serves to give a better understanding of the anthropological magnitude of climate change effects, the actual migration patterns of refugees might be more complex than this specific estimation. There is uncertainty with regard to the frequency and intensity of climate change effects that makes it impossible to have an accurate future estimation. Given past observations with respect to migration patters, it is most likely that there will be internally displaced people being challenged by climate change effects.

With regards to the preparedness of the world, I believe there is a lot more to be done, starting by creating and rigorously implementing domestic policies and national adaption plans that will consider the implications in various sectors of the economy. The Paris 2015 Climate Change Agreement is hopefully the first step towards the global effort to increase the adaption and resilience of people and the environment.

Describe a particular area of climate change that you are concentrating upon with research or a grassroots development project.

Water resource management in vulnerable communities.

Notes

1. IPCC, Fifth Assessment Report, 2014.
2. "The economics of climate change in Latin America and the Caribbean Paradoxes and challenges of sustainable development," Economic

Commission for Latin America and the Caribbean, ECLAC (2012a), Sustainable Development 20 Years on from the Earth Summit: Progress, gaps and strategic guidelines for Latin America and the Caribbean (LC/L.3346/Rev.1).
3. FBDS, "Climate change and extreme events in Brazil" Lloyds. http://www.lloyds.com/NewsCentre/360riskinsight/Thedebateonclimatechange.
4. "Climate Change and Health Country Profile, Brazil" WHO, 2015.
5. Atlas Brasileiro de Desastres Naturais 1991 a 2010: Volume Brasil. Centro Universitário de Estudos e Pesquisas sobre Desastres. Universidade Federal de Santa Catarina; Florianópolis/SC, Brasil: 2012. [(accessed on 25 July 2014)]. CEPED-UFSC. Available online: http://150.162.127.14:8080/atlas/Brasil%20Rev.pdf.
6. CPTEC/INPE.
7. "Climate change and population migration in Brazil's Northeast: scenarios for 2025–2050" Alisson F. Barbieri, Edson Domingues, Bernardo L. Queiroz, Ricardo M. Ruiz, José I. Rigotti, José A. M. Carvalho, Marco F. Resende, Springer Science+Business Media, LLC 2010.
8. Tercio Ambrizzi, University of São Paulo's Institute of Astronomy, Geophysics and Atmospheric Sciences (IAG-USP).
9. "Researchers present projections for extreme weather events in Southeast Brazil" October 26, 2016, by Diego Freire, Agência FAPESP and Tercio Ambrizzi, University of São Paulo's Institute of Astronomy, Geophysics and Atmospheric Sciences (IAG-USP).
10. Ibid.
11. IPCC "Climate Change 2014: Impacts, Adaptation, and Vulnerability" 2014.
12. "Climate Change and Health Country Profile, Brazil" WHO, 2015.
13. Climate Action Tracker, Brazil, 2016. http://climateactiontracker.org/countries/brazil.html.
14. Ozwaldo Cruz Foundation, 2016. https://portal.fiocruz.br/en/content/foundation.
15. Ibid.
16. GFDRR, Country Program Update, Ecuador, May 2014.
17. "Amid Predictions of More Extreme Weather Events, General Assembly Urges Quick, Coordinated Humanitarian Action to Address El Niño Phenomenon" GA/11851, 2 November 2016, United Nations General Assembly, Plenary, Seventy-first Session, 39th Meeting.

18. Deborah Hines, Verónica Alvarado and María Victoria Chiriboga, "Enhancing climate resilience in Ecuador's Pichincha Province and the Jubones River Basin", Hunger, Nutrition, Climate Justice: A New Dialogue: Putting People at the Heart of Global Development, 2013.
19. "Climate change and agricultural workers' health in Ecuador: occupational exposure to UV radiation and hot environments" Raul Harari Arjona, Jessika Piñeiros, Marcelo Ayabaca, and Florencia Harari Freire, Ann Ist Super Sanità, Vol. 52, No. 3: 368–373, 2016.
20. Cueva P., Yepez J., (Eds). Cancer Epidemiology in Quito 2006–2010. National Cancer Registry. Quito: SOLCA, 2014.
21. Casale M.C., Todaro G. La patología cutánea di origine professionale. INAIL, Sovrintendenza Medica Generale. Italia: INAIL, 1999.
22. Siani A.M., Casale G.R., Sisto R., Colosimo A., Lang C.A., Kimlin M.G. Occupational exposure to solar ultraviolet radiation of vineyard workers in Tuscany (Italy). Photochem. Photobiol. 2011(87): 925–934.
23. Deborah Hines, Verónica Alvarado and María Victoria Chiriboga, "Enhancing climate resilience in Ecuador's Pichincha Province and the Jubones River Basin" Hunger, Nutrition, Climate Justice: A New Dialogue: Putting People at the Heart of Global Development, 2013.
24. "Argentina Country Note on Climate Change Aspects in Agriculture" World Bank, December 2009. www.worldbank.org/lacagccnotes.
25. Assessment of Disaster Risk Management Strategies in Argentina, Government of Argentina, 2011.
26. http://www.emdat.be/Database/CountryProfile/countryprofile.php?disgroup=natural&country=arg&period=1999$2008.
27. Rusticucci M., Kysel'y J., Almeira G. Lhotka O. Long-term variability of heat waves in Argentina and recurrence probability of the severe 2008 heat wave in Buenos Aires. WCRP Conference for Latin America and the Caribbean, Montevideo, 2014. http://www.cima.fcen.uba.ar/WCRP/docs/pdf/Abstract_M-Rusticucci.pdf.
28. http://www.ambiente.gov.ar/archivos/web/UCC/File/poster_vulnerabilidad_produccion_agricola.pdf.

4

Southeast Asia: Thailand, Myanmar, Japan

> *"Ultimately the decision to save the environment must come from the human heart."*
> The Dalai Lama

Southeast Asia, including China, has a population of 2.1 billion (2018) and is projected to face some of the most severe climate change effects in the twenty-first century. The countries of the region include Brunei, Cambodia, China, Indonesia, Laos, Malaysia, Myanmar, the Philippines, Singapore, Taiwan, Thailand and Vietnam. The climate is tropical, warm and wet, and notable for tropical storms, thus contributing to the significant flood risk that is paramount for many Southeast Asian nations along with the destructive eventuality of sea level rise and saltwater intrusion.

Several reports have noted the considerable climate related challenges facing the region and high urgency for sustained mitigation and adaption strategies. "The region is highly exposed to extreme weather events, and both warming and extreme events—high temperatures and heavy precipitation—are projected to increase in future decades."[1] Agriculture, which accounts for more than 10% of GDP in most countries in the region, is highly sensitive to climate effects, and persistent poverty in rural areas, low levels of education, spatial isolation and neglect by policymakers, can amplify that impact.[2] Vulnerability is particularly high in rural areas, which are home to more than three-quarters of the region's poor people, many of them smallholders or subsistence farmers.[3] According to the IPCC, Asia has a number of significant vulnerabilities to climate change induced crises. "People living in low-lying coastal zones and flood plains are probably most at risk from climate change impacts in Asia. Half of Asia's urban population lives in these areas. Asia has more than 90% of the global population exposed to tropical cyclones. Asia is predominantly agrarian, with 58% of its population living in rural areas, of which 81% are dependent on agriculture for their livelihoods. Rural poverty in parts of Asia could be exacerbated due to negative impacts from climate change on rice production, and a general increase in food prices and the cost of living (high confidence)."[4]

Thailand

Thailand is a major agricultural nation and leading global rice exporter with a population of 70 million (2018), a territory of 514,000 km² and a coastline of 3219 km. The GDP is USD 390 million and the HDI rank is 93. The level of CO_2 emissions in 2016 was 0.77%. The major climate change implications to face the country are drought in the northeast, persistent flooding, coastal flooding—particularly affecting the metropolis of Bangkok—and extreme heat episodes. All of these symptoms will bring forth considerable health, economic, infrastructure and internal environmental migration challenges.

Analysis of climate data and trends indicate some likely environmental scenarios. "Major climate models indicate a temperature rise for the whole country of Thailand, particularly the central plain and lower North-eastern region." Projections for the increase of mean temperatures vary between 0.4 and 4.0 °C in the next 100 years.[5] In addition, the number of warm days >35 °C daily mean temperature) per year is expected to increase, particularly in the Chao Phraya River basin, central plain, and lower northern regions, meaning an extension of the summer/hot period (with maximum daily temperature > 35 °C) of 2–3 months on average. The northeastern, central, and southern regions are expected to have hot periods extended to 5–6 months, by the end of the century, while the northern region is expected to extend to 3–4 months.[6] The duration of the cold period (with cold days, temperature < 16 °C) in the north and northeast will shorten after mid-century from a current 2–2.5 months to 1–2.5 months.[7] This data confirms numerous reports about worsening factors contributing to enhanced drought conditions in the north, and in particular the northeast region.

Flooding and Drought

Heavy rainfall and flooding have been constants in Thai history, though the incidence and severity is expected to increase along with drought episodes in the northeast. In 2009, heavy rains caused widespread flooding across 11 provinces in southern Thailand. An estimated 330,300 households were affected and the authorities reported 21 fatalities. The Thai government upgraded the disaster management response level to three (large-scale disaster) and set up disaster response centers in Surat Thani and Songkla. In 2011, Thailand was devastated by the worst flooding in decades. More than a million citizens were impacted by the flooding which continued for three months. The 2011 Chao Phraya River flood that caused USD 45 billion of damage was a somber reminder of the devastating human and financial toll of flooding in the developing country. In 2016–2017 persistent flooding was reported across wide areas of the country involving 12 provinces including Chumphon, Surat Thani, Nakhon Si Thammarat, Phatthalung, Songkhla, Pattani, Yala, Narathiwat, Phuket, Krabi, Trang and Satun. Approximately 1.8 million people and 590,000 families were affected. Moreover, infrastructure damage struck 4314 roads, 348 bridges, 270 drains, 126 weirs, two reservoirs, 70 government offices and 2336 schools. In Nakhon Si Thammarat, five people died and 260,000 residents in 161 tambons of 23 districts were devastated by the flooding. In Surat Thani, 87,000 villagers were affected with many displaced by the emergency that left 16 out of 19 districts declared flood-affected zones, including Muang, Chai Buri, Tha Chana, Tha Chang, Koh Samui, Ban Na San, Phrasaeng, Ban Na Doem, Chaiya, Khiri Ratthanikhom, Kanchanadit, Don Sak, Phunphin, Wiang Sa, Koh Phangan and Khian Sa.

One of the main climate scenarios that will hit Thailand with increasing ferocity is rising sea levels and the attendant flooding that will follow. The sea level in the gulf of Thailand has risen about 3–5 mm per year between 1993 and 2008 compared to a global average of 1.7 (±0.5) mm per year.[8] The effects of sea level rise overlaid with land subsidence may mean up to 25 mm per year of net sea level rise in some areas such as the larger Bangkok metropolitan area or the river mouths in the gulf of Thailand.[9] The mean sea level of the Andaman coast in Krabi province is expected to rise by about 1 cm annually over the next 25 years, with shoreline shifts of between 10 and 35 m.[10]

Drought is a constant in Thai history and is expected to increase in severity and duration in the twenty-first century. There are 14 million Thais employed in the agriculture sector. Climate change impacts on agriculture will have momentous effects upon food security, incomes, health and development. Farmers and food production are extremely weather dependent. Thus variations in weather, heat and water cycles can have devastating effects upon the agriculture sector and the millions of Thais, many working at subsistence level, who depend upon farming. The noble Royal Rainmaking Project, guided by the late King of Thailand, who was an expert in agriculture, was utilized effectively in many instances to combat regional drought in the north.

A recent analysis of drought conditions in the country noted several issues of concern. "It is commonly understood that most droughts are caused by climate change that leads to sub-average rainfall or aberrations in seasonal rainfall patterns, reducing accumulated rainfall from year to year. According to data from the Land Development Department, areas of permanent droughts in Thailand still covers as much as 40% of total agricultural areas, suggesting that apart from the decline in rainfall, there are the other factors that contribute importantly to droughts in Thailand as well, such as water consumption behaviors, agricultural and industrial land development, and population growth. To make matters worse, in 2014 and 2015, annual accumulated rainfall in Thailand dipped below the 30-year average (1981–2010) for two consecutive years, causing a significant depletion of water supply across various reservoirs in 2016, leading to the current situation."[11]

The Chao Phraya River is a vital waterway for the city of Bangkok and commercial activity that benefits the entire nation. Flooding in the lower Chao Phraya floodplain has been a constant for centuries, yet with sea level rise and other effects of climate change the level of flooding will inevitably increase significantly. Bangkok itself has been identified as one of the world's most threatened coastal cities from sea level rise and saltwater intrusion. According to a World Bank report on flood risks in Asia: "The huge volume of water (over 31,000 million cubic meters for a 1-in-30-year event) associated with these floods can take months to drain as Bangkok acts as a 'bottleneck' to the discharge reaching the Gulf of Thailand. This can leave many areas flooded for more than 30 days. The flood protection schemes for Bangkok are generally designed to protect against 1-in-30-year floods. There are steep increases in persons affected as inundation levels exceed the 2008 1-in-30-year design standards. For example, the number of persons who would be affected by 1-in-100-year flood event in 2008 is nearly double the number affected by a 1-in-30-year flood. In a C2050-LS-SRSS-A1FI-T30 scenario, almost 1 million people in Bangkok and Samut Prakarn would be impacted by floods. The impact will be profound for people living on the lower floors of residential buildings."[12]

In 2010, Thailand experienced the most severe drought in two decades, which caused water levels in the vital Mekong River to reach its lowest level in 50 years. Overall, 7.6 million people in 59 provinces were impacted. There are approximately 24 million Thais living in the lower Mekong basin. The fisheries of the Mekong are a USD 2 billion annual market. Thai officials are concerned that exponentially increasing drought conditions in the twenty-first century, which are forecasted according to data projections, will place severe human and financial burdens on the population and divert valuable resources away from other priority issues.

Extreme Weather

Extreme weather events are symptomatic of climate change. "As evidenced by the recent drought, climate change is an important issue for Thailand in both the medium and long term. Floods, droughts and tropical storms—which cause numerous natural disasters annually—will only multiply in frequency and intensity. Major climate-induced changes could have severe negative impacts on Thai food production, particularly rice. Recent scientific studies argue that climate change is a significant factor in the increasing frequency and intensity of tropical storms during the last 35 years."[13] Moreover, recent data suggests that there is a greater than 66% probability that sea-surface warming will cause tropical storms to become more intense and heighten storm surges during this century. "These surges will cause more damaging floods in coastal areas and low-lying areas."[14] Studies have predicted more intense bursts of precipitation and El-Niño and La-Niña events, which have been arising every four to ten years in Thailand.[15] "Other studies report that numerous climate extreme events such as more recurrent and more intense droughts, floods, cyclones, and extreme rainfall events pose increasing and, oftentimes, limitless threats to environment, water resources and agricultural production will affect Thailand."[16,17]

Health, Human Development and Food Security

There is certainty among climate scientists and health experts that climate change will have numerable and deeply consequential impacts upon human health. "Under a high emissions scenario, and without large investments in adaptation, an average of 2.4 million people are projected to be affected by flooding due to sea level rise every year between 2070 and 2100. Under a high emissions scenario heat-related deaths in the elderly (65+ years) are projected to increase to about 58 deaths per 100,000 by 2080 compared to the estimated baseline of about 3 deaths per 100,000 annually between 1961 and 1990. By 2070, approximately 71 million people are projected to be at risk of malaria assuming a high emissions scenario. If emissions decrease rapidly, projections indicate this number could decrease slightly to about 66 million."[18]

In 2016, the agricultural sector contributed about 11% of the GNP and employed more than 12 million Thais. Agricultural production is mainly generated by small-scale farms. Hence weather extremes can have an injurious effect on both farm production and the socio-economic well-being of millions of farmers whose profit margins are extremely thin. Between 1989 and 2010, droughts and floods affected 2–27% of total agriculture lands and resulted in crop losses ranging from THB 1000–17,000 million (EUR 25–425 million) per annum.[19]

The impacts of climate change upon food security and production will be substantial. This is primarily due to the projected increases in northeastern drought, heat waves, heavy rains and coastal flooding. "Climate change will have varying effects on Thai crops. Heavy rain may damage the roots of cassava plants in the north, while a decrease in rain might damage cane sugar and rice in the central region. Temperature and quality changes of water might lead to a reduction in the viability of livestock due to heat stress, survival rates of newborn animals, and immune system impacts."[20] Climate change has had a devastating effect upon rice cultivation and harvests, which represents a particularly difficult situation for Thailand as the second largest rice exporter in the world. A study by Okayama University in Japan found that grain yield declines when the average daily temperature exceeds 29 °C (84 °F) and grain quality continues to decline linearly as temperatures rise.[21] In 2015–2016, Thai rice production declined by 16% due to weather conditions. Overall, Thailand lost 6.1 million metric tons of agricultural products worth THB 15.5 billion between January 2015 and April 2016.[22]

Climate Change Refugees

Although Thailand is not expected to generate external climate refugee flows, the number of citizens forced to relocate from areas impacted by drought, flooding and coastal flooding is expected to increase in successive decades. The substantial flooding that is projected for Bangkok will force thousands of Thais, mainly low-income citizens to move inland. As noted, climate change poses a disproportionate burden upon the poor, elderly and health challenged. The Klong Toey slums in Bangkok, which are situated along the low-elevation canal areas of the city have a history of flooding, inadequate and unsafe dwellings and extreme poverty. Approximately 20% of Bangkok residents live in illegal squatter settlements. In Klong Toey, the average household income is half the national average. Bangkok, has been sinking at the approximate rate of 10 cm annually. Land subsidence, saltwater intrusion and sea level rise endanger the city to the point of devastating submersion within two decades according to Smith Dharmasaroja, former chairperson of the Thailand government Committee of National Disaster Warning Administration. Discussions have been initiated to construct a THB100 billion (USD 3 billion) flood prevention wall to protect Bangkok that will be 80 km long and situated 300 m offshore. A buffer of mangrove forest would supplement the protection wall to further shield the coast from flooding.

Innovation and Adaption Strategies

1. Thailand signed the Paris Agreement in 2016, but has not yet ratified the agreement;
2. Thailand Climate Change Master Plan 2015–2050;
3. National Strategic Plan on Climate Change Management (2012);
4. establishment of Thai Greenhouse Gas Management Organization (2008);
5. Thailand Established Climate Change Management and Coordination Division (2004);
6. Thailand ratified the Kyoto Accord (2002);
7. Thailand signed the UNFCCC in 1992;
8. Thailand Energy Conservation and Promotion Act (1992).

Additional important climate initiatives include the following:

1. The Bureau of Royal Rainmaking (BRR) has established an operation center in the north to combat the persisting drought and haze. The center has been set up in the Muang district of Chiang Mai province with the goal of minimizing drought impact.
2. A national focal point for climate change in the Ministry of Health was established.
3. The National Communication submitted to UNFCCC includes health implications of climate change mitigation policies.
4. Thailand has initiated plans to build institutional and technical capacities to address climate change and health.

5. Thailand initiated a national assessment of climate change impacts, vulnerability and adaption for health.
6. An Integrated Disease Surveillance and Response (IDSR) system, including development of early warning and response systems for climate-sensitive health risks have been initiated.

Myanmar

As Myanmar emerges from three decades of oppressive military dictatorship and widespread poverty, one of the many development challenges is to launch effective climate change protection and mitigation policies. The country has a population, primarily rural, of 55 million (2017), a land area of 261,969 square miles (678,500 km^2) and coastline of 2228 km (1385 miles). The GDP is USD 62 billion (2015) and the country has an HDI ranking of 145. According to the Global Climate Risk Index 2015, Myanmar is the second nation most impacted by climate change in the world.

Myanmar has abundant water resources and fertile land. Approximately 70% of the population is engaged in agriculture, which renders climate change impacts more serious and potentially destructive to the economic, health and development of the impoverished nation. Rural, impoverished communities will be particularly challenged. Approximately 66% of the population live in rural areas and 33% live in urban areas such as the capital city of Yangon. The most significant climate change issues facing the country include: sea level rise, drought, a decreasing southwest monsoon duration, increasing cyclone seasons, increasing cyclone incidents, extreme rain spells and extreme heat episodes. A recent Myanmar government report stated that, "There is an urgent need for Myanmar's communities and economic sectors to adapt to climate change and variability. This is because actual and potential impacts are becoming more evident, including: (1) increased temperatures, (2) rainfall variability; and (3) more frequent, intense and widespread extreme weather events such as cyclones, floods, intense rains, droughts and high temperatures. Predicted climate change in Myanmar is expected to have negative impacts on the entire socioeconomic functioning of the country."[23]

Flooding and Drought

Flooding and drought are chronic problems facing Myanmar. Climate projections over the next century indicate an increase in the intensity and duration of this weather pattern. A government report noted the following climate change and weather threats to the population: storm surge flooding as a result of cyclones and storms in coastal areas; river floods in delta areas; flash floods, localized floods in urban areas as a result of inter alia cloudburst; damage to coastal areas, for example mangroves and river ecosystems; severely flooded or inundated land (every 2 years approximately 2 million ha of land is flooded and 3.25 million ha is moderately inundated); intense runoff and soil erosion (particularly during La Niña periods); damage to riverbanks and irrigation systems; entire villages lost, including loss of lives and livelihoods; heat waves; and reduced water availability and aggravated drought events.[24]

In 2015, 146 villages experienced water shortages during the summer period, and in 2016 more than 300 villages faced similar conditions. In 2016, Myanmar also faced severe after effects from El Niño that included heat extremes, parched soil and water scarcity in many regions. The drought conditions impacted economic conditions, health and employment.

Extreme Weather

The country routinely experiences extreme weather events that are expected to be worsened by climate change impacts. The population is concentrated in the Ayeyarwaddy basin, which is susceptible to cyclones, tropical storms and flooding. "From 1887 to 2005, 1248 tropical storms formed in the Bay of Bengal. Eighty of these storms (6.4% of the total) reached Myanmar's coastline. Recent cyclones of note include Cyclone Mala (2006), Nargis (2008) and Giri (2010).[25] In 2008, Cyclone Nargis struck the coast of Myanmar and caused massive loss of life and extensive economic damages. It was the worst Cyclone to ever reach the country. The main impacts included: (1) extensive damage to mangroves, agricultural land, houses and utility infrastructures; (2) salt-water intrusion into agricultural lands and freshwater sources causing economic, social and environmental damage; (3) loss of livelihoods and homes (3.2 million people affected), including 138,373 deaths; and (4) damages of ~US $4.1 billion."[26]

A climate assessment report by the government of Myanmar noted the variety and severity of extreme weather events that routinely effect the country: "(1) cyclones/strong winds; (2) flood/storm surge; (3) intense rains; (4) extreme high temperatures; and (5) drought." Drought is the most severe weather event in the country, followed by extreme day temperatures, cyclones, strong winds, intense rain and flood/storm surges.[27] Some of the more severe impacts from extreme weather include: damage to agricultural land and crops; windthrow of trees; damage to vulnerable ecosystems, for example coral reefs and mangrove/coastal forests; local rice/crop landraces (germplasm) damaged and/or lost; loss of lives and livelihoods; and displaced ground and surface freshwater supplies by saline water. Indirect limiting of drinking-water and irrigation-water supply.[28]

Health, Human Development and Food Security

It is certain that inland and coastal flooding, destructive sea level rise and saltwater intrusion, drought, extreme heat, storms and cyclones will increasingly threaten the population of the country. With these climate challenges come severe health risks to a population that is largely rural (70%) and has a high poverty ratio of 26%. In Myanmar, poverty stricken and marginalized communities, such as indigenous hill tribes, will face particularly devastating consequences. The climate change risks to health and development are multiple and will be sustained over many decades. "Increasing temperatures and erratic precipitation patterns will create favorable conditions for the spread of infectious diseases. Additional effects of increasing temperatures on human health, including inter alia heat stress, heat exhaustion and dehydration. Additionally, extreme temperatures will aggravate the effects of individuals suffering with cardiorespiratory diseases. Air pollution-related health problems will also be affected as air pollution levels increase from a combination of high emissions and unfavorable weather."[29] Higher temperatures will also reduce the development time for pathogens and thereby increase transmission rates, for example mosquito-borne diseases such as malaria and dengue will increase.[30] Furthermore, pathogen distribution will increase in range as vectors harboring parasites infest highland areas (e.g. Shan State), which at present are too cold for vector insects.[31] Increases in flooding events and storm surges will affect freshwater sources as they become contaminated by rising flood water levels. Furthermore, rising sea levels will result in fresh groundwater resources being displaced with saltwater. "In addition, increases in occurrence and severity of droughts will decrease water availability and water quality."[32]

Agriculture is the main economic industry in the country and employs approximately 65% of the workforce. Moreover, about 30% of the rural population is landless and has no stable income source other than work as manual and seasonal laborers in the agriculture sector. The primary crops cultivated include rice, corn, wheat, millet, cotton and sugarcane. Livestock is important to the economy as well as jute and rubber. All of these products are extremely sensitive to fluctuations in climate, thus have a subsidiary effect on socio-economic conditions. The Ayeyarwady delta is an important agricultural region yet increasingly threatened by flooding and saltwater intrusion. The most vulnerable communities in Myanmar occur in all three agro-ecological zones, namely the hilly, dry and coastal zones, and are made up of mainly community group members situated in high risk areas and participating in vulnerable livelihood strategies, for example farmers, woodcutters, fisher folk, grocery merchants, casual workers, homemakers and retailers.[33] "Approximately 70% of the rural population is dependent upon rain-fed agriculture, livestock and fishery and forest resources. The livelihood of rural communities and the productivity of the agricultural sector as a whole are therefore largely influenced by climate conditions in these areas. The economy of Myanmar and its society is therefore highly sensitive and vulnerable to climate change, climate variability and natural disasters."[34]

Climate Change Refugees

The weather patterns in the country and the vulnerability to drought and flooding of rural communities as well as flooding scenarios in coastal communities combine to create internal environmental refugee pressures. Although external climate refugee flows are not anticipated in the near future, the impact of domestic challenges will push some citizens to migrate to neighboring countries such as Thailand for better economic opportunities. Moreover, as a developing country emerging from decades of regressive military rule, it will be difficult for the national democratic government to meet the financial and policy demands of the affected climate refugee communities.

Innovation and Adaption Strategies

1. Myanmar signed the Paris Climate Change Accord (2016)
2. The Disaster Management Law (2013) law is designed to initiate disaster management programs; coordinate with government departments on responses to disaster management; re-build affected areas; develop and provide restoration services in health and development to affected individuals; and bring about better living conditions for victims.
3. National Bio-diversity Strategy and Action Plan (2012).
4. Myanmar joined the UNREDD program (UN collaborative initiative on Reducing Emissions from Deforestation and forest Degradation in developing countries) in 2011.
5. National Environment and Health Action Plan (2010) addresses several important issues including: water security, air quality, environmental health systems, climate change, ozone depletion and ecosystem changes.
6. The National Sustainable Development Strategy (2009) includes three important components: the sustainable management of natural resources, economic development and sustainable social development.
7. Myanmar ratified the Kyoto Protocol (as a non-Annex I country) in 2003.
8. Myanmar ratified the UNFCCC in 1994.
9. By 2030, Myanmar's Permanent Forest Estate (PFE) goal is to increase the amount of forested land that includes Reserved Forest (RF) and Protected Public Forest (PPF), which will constitute 30% of the total national land area. Protected Area Systems (PAS) will be 10% of the total country land area.

Japan

Japan has a population of 127 million (2017) and a land area of 377, 930 square miles with a coastline of 18,486 miles. It is the third largest economy in the world with a GDP of USD 4.7 trillion, which will assist the country in addressing substantial climate change adversities and its HDI rank is 20. Japan's CO_2 emissions are the fifth highest globally at 3.6%. Its main climate change threats include sea level rise, saltwater intrusion, flooding, extreme weather events and health threats to a rapidly aging population. In 2017, approximately 34% of the population was over the age of 60, 26% over age 65 and 13% above age 75. The low birthrate of 1.41 (2017) is well below a population replacement rate of 2.1% thus leading to the situation of a rapidly aging population who will require substantial care and policy measures to mitigate the effects of climate change induced health problems.

Flooding and Drought

As an island nation in the crosscurrent of storms, hurricanes, typhoons, substantial sea level rise as well as rising extreme heat episodes, it is clear that Japan will be impacted in dramatic ways by environmental change. Projections for the country include: a temperature rise of 2–3 °C over the next 100 years; an increase in the number of extreme hot days; an increase in precipitation of more than 10% over the twenty-first century and a projected sea level rise of 5 mm per year throughout the twenty-first century. Approximately 10% of the world's population live in coastal areas that are less than 10 m above sea level with risk ratios for human and property devastation ranging from mild to severe. One of the most affected countries is Japan.

> Between 1994 and 2004, about one-third of the 1562 flood disasters, half of the 120,000 people killed and 98% of the 2 million people affected by flood disasters were in Asia, where there are highly populated areas in the flood plains of major rivers (e.g. Ganges–Brahmaputra, Mekong and Yangtze) and in cyclone-prone coastal regions (e.g. Bay of Bengal, South China Sea, Japan and the Philippines).[35]

There are several influencing factors that contribute to rising sea levels. These include thermal expansion of sea water; growth/decay of land-based glaciers and ice caps; growth/decay of ice sheets, such as the Greenland Ice Sheet (GIS) and West Antarctica Ice Sheet (WAIS); terrestrial water storage—that is dam reservoirs, lakes and so on—and depletion of groundwater; tectonic movement including ground subsidence/uplift associated with earthquakes; ground subsidence/uplift due to compaction of the ground, pumping up of ground water; changes in ocean currents; changes in atmospheric pressure and tide, tsunami, storm surges and waves.[36]

Although Japan is a the third largest global economy and has substantial mitigation and adaption capacity, the human and infrastructure damage over the twenty-first century from sea level rise, increased tropical storms and the attendant flooding and saltwater intrusion will be immense. According to recent analysis, "This study indicated the following results assuming that the typhoon (tropical cyclone) becomes 1.3 times stronger than the Ise Bay Typhoon in 1959; the inundation areas and people at risk will be 63–72 km^2 and 300–350 thousand, respectively, and economic damage will amount to 1.8–2.3 trillion JPY. This study also showed that if sea-level rise and intensified storm surges are superposed, the current return period of 1/several hundred years for a high water level will be shortened to 1/several decades. The significance of this result is that the current design heights of seawalls might be insufficient for the higher sea level expected in the future."[37] Moreover, the possibilities of floods, which exceed final river maintenance targets, are projected to be 1.8 to 4.4 times higher than current levels in Japan.[38]

Drought is a less serious eventuality for Japan, though still a factor in southern regions where water levels have fallen. Moreover, the projections of a temperature increase of 2–3 °C over the next century and 4 °C in the Sea of Okhotsk area enhance drought conditions. Accordingly, the number of extreme hot days with temperatures surpassing 35 °C is expected to increase.

Extreme Weather

Projections for the country indicate that the frequency and duration of extreme weather events will increase. Studies by Oouchi et al. (2006)[39]

and Kitoh et al. (2009)[40] hypothesize that the number of stronger typhoons will increase, while the total number of typhoons may decrease. The risk facing Japan is that water related environmental adversity will be elevated to extreme degrees. Moreover, with a coastline of 18,486 miles, damage to coastal communities from typhoons, sea level rise, flooding and saltwater intrusion will be extensive. In Japan, "policies and facilities for coastal protection have been developed. However, climate change would bring about storm surges, waves, and floods higher than the current protection standards."[41]

Health, Human Development and Food Security

The IPCC noted in the Fourth Assessment Report that Japan will encounter a likely scenario of more frequent heat waves, intense rain and stronger typhoons, which will significantly impact health, development, infrastructure and agriculture. Furthermore, it has been projected that the seas around Japan could rise as much as 40–60 cm by 2100, thereby placing millions of citizens at risk and costing billions in property, infrastructure and associated health care. "In the coastal zone, an estimated 1.3 million people could be vulnerable to storm surge flooding, especially in western Japan. In terms of human health, it is predicted that the number of people at low risk of death from heat stress could double by 2100, and those at high risk may increase fivefold."[42]

The challenge with climate change impacts relates to deep and sustained burdens at two levels. First, there are the extreme effects faced by vulnerable groups within society, namely the elderly, health compromised, disabled and impoverished. Although Japan has a relatively high standard of living there are still segments in society who face socio-economic barriers and thus would be unable to respond adequately to health and economic costs associated with climate change mitigation. Second, infrastructure damages in all segments of the country will impose severe costs upon governments and citizens. A report by the Japanese government summarized the multiple health threats from climate change that are current and will continue to rise. "Increase in heat-wave intensity and heat stress, putting vulnerable populations, such as the aged, at risk;

increased likelihood of infectious and vector and water-borne diseases; expansion of dengue fever into Hokkaido; increased allergies and allergy-related diseases; increased cost of living and protection from more extreme weather events; 67 to 70% increase in wind-related losses from more intense typhoons."[43]

With respect to food production and agriculture, Japan will experience notable crop damage and agricultural disruption due to environmental factors. Moreover, rice, which is a staple of the Japanese diet and a leading crop, will experience significant downward pressures. Japanese climate expert Mimura has highlighted some of the impacts that will confront the country. "Japan will have to cope with a range of impacts that could simultaneously: affect the stability of its most important food supply, rice; increase the severity of natural disasters, like flooding; alter and undermine some of its key natural ecosystems; place even greater pressure on communities and a health system already struggling with its ageing population; and cause increased deaths from heat stroke and communicable diseases."[44]

It is estimated that the average temperature in Japan will rise between 2.1 and 4 °C by the end of the twenty-first century, thus contributing to extreme weather events and pronounced damage to rice production. In 2016–2017, Japan was the 11th largest global producer of rice at 7,780,000 metric tons. "A 2010 survey noted the following negative yield results on the quality and crop situation of irrigated rice: Japan total—17.6%; Chugoku/Shikoku—18.5%; Hokoriku—40.1%; Kinki—32.4%; Tokai—32.5%; Kanto—14.9% and Tohoku—14.1%."[45] "Other impacts from climate change upon agriculture include an estimated 40% decrease in rice yields in central and southern Japan, potential northern shift of some fish species and changes in the fish abundance and diversity."[46]

Climate Change Refugees

Japan is not expected to generate external environmental refugee flows. Moreover, considerable national wealth and a high standard of living will preclude this eventuality. However, a serious internal environmental migration flow is inevitable from coastal areas impacted by sea level rise and large-scale flooding. In 2011, 230,000 people were displaced following the Tohoku earthquake and the tsunami disaster. An estimated 20,000 citizens were killed, economic damages from the tsunami were estimated at USD 300 billion and approximately 23,600 ha of farmland, mostly rice fields, were severely eroded. The environmental projections for Japan include a rise in tropical storm activity. Moreover, climate scientists report that earthquakes can be exacerbated by climate change. Due to its geographical position, as an archipelago situated on the Ring of Fire, Japan has always been subject to natural risks such as earthquakes, tsunamis, typhoons and volcano eruptions, thus for the Japanese people, nature is inherently risky and they have to coexist with that risk—this is called the "saigai" conception.[47] In 2015, 100,000 people were displaced from their homes by massive rainfall levels of up to 500 mm in one day that caused heavy flooding. As climate change induced extreme weather events and flooding mount in severity, there are expectations that high numbers of environmental migrant flows will emerge.

City Profile: Tokyo

The population of Tokyo is 13.8 million (2017) and 38 million in the greater metropolitan region with an area of 2188 km^2. Tokyo is situated at the head of Tokyo Bay on the Kanto Plain. Tokyo has 23 wards, 26 smaller cities and several towns and villages that stretch 55 miles from east to west and 15 miles north to south. The most severe and pressing environmental threat facing Tokyo is coastal flooding due to sea level rise and the inter-related problem of saltwater intrusion.

A scientific report addressing sea level rise in major coastal cities, including Tokyo, concluded with dire projections. "If a tropical cyclone (typhoon) becomes stronger, higher incident waves will occur, resulting in an even higher run-up. Isobe (2013) shows a comparison of run-up heights on a model seawall in Tokyo Bay, calculated for the conditions with and without sea-level rise and stronger typhoons. As a model tropical cyclone, Ise Bay Typhoon in 1959 with a pressure depression of 70 hPa was taken, because it caused a historically high storm surge of 3.4 m in Japan. If a 60 cm sea-level rise and a stronger typhoon with 10% increase in the pressure depression are assumed, the difference in the maximum run-up height with and without the assumptions is more than three times larger than the sea-level rise of 60 cm. This means that, for the planning of a countermeasure, it is not enough to raise the seawalls by the amount of projected sea-level rise, but necessary to take into account the increased run-up height."[48,49]

Innovation and Adaption Strategies

1. Japan ratified the Paris Agreement in 2016.
2. December 2015 Japan formulated its Policy for Global Warming Prevention Measures Based on the Paris Agreement.
3. In 2016 the cabinet approved both the Plan for Global Warming Countermeasures based on the Act on Promotion of Global Warming Countermeasures and the Government Action Plan. An important goal is the 26% reduction of GHG emissions by fiscal year (FY) 2030 and 80% reduction by FY 2050.
4. In 2016 the Act on Promotion of Global Warming Countermeasures was revised.
5. In 2015 Japan submitted its INDC which includes an emissions reduction target of 26% below 2013 emission levels by 2030, equivalent to 18% below 1990 levels by 2030.
6. At COP 21, Japan pledged to provide, in 2020, approximately JPY 1.3 trillion of public and private climate finance (1.3 times up from the current level) to developing countries.
7. Japan Center for Climate Change Action (JCCCA)—local centers are assigned to 47 prefectures and 8 cities as of December 2013.
8. Japan is promoting the diffusion of combined heat and power such as household fuel cells— "Ene-farms." The goal is to introduce 5.3 million household fuel cells—"Ene-farms"—into the market by 2030.
9. The government will extensively promote energy-efficient vehicles— namely hybrid vehicles, electric vehicles, plug-in hybrid vehicles, fuel-cell vehicles, clean diesel vehicles and compressed natural gas vehicles—with a goal to increase the share of these vehicles in new car sales from 50 to 70% by 2030.

Interview

Dr. Apichart Anukularmphai is an agricultural and water resource engineer by training and has been working in this field for more than 50 years. He is an expert in integrated water resource management.

Can you describe your work related to climate change?

I work mainly on water resources management in relation to flood and drought, hence I have to take climate change into consideration as well. While a lot of studies and research is on-going, from a practical point of view the introduction of risk management into project planning processes is a suitable option.

What concerns you the most about climate change in your country?

Climate change affects the frequency and severity of flood and drought from the normal average. In our case, we were faced with severe flooding in 2011 and severe drought in 2016, so it is important to review and adopt new approaches in managing water resources by taking climate change into account.

What concerns you the most about climate change globally?

In a global context, we should try to reduce CO_2 emissions and use "green technologies" in industry as well as agriculture. However, it seems to be difficult to achieve global agreement and full cooperation due to conflicts of interest among key players.

Can you describe a specific sustainability project in your area or country that is contributing to a "greener world" and lower carbon footprint?

In Thailand, the government is promoting alternative energy to reduce fossil energy, and the industrial sector is engaging in the introduction of green technologies. Also, work is on-going to establish a protocol for carbon-credit trade.

What steps could your government take to mitigate the growing threats from climate change?

The Thai government has pledged to reduce carbon emissions with a set target and time frame, but more work has to be undertaken.

What is your reaction to the UNIPCC's accepted calculation by Prof. Myers of a minimum of 200 million climate change refugees by 2050 and the world's readiness to cope with this humanitarian and environmental challenge?

Though my area of interest is closely related to climate change, I do not work with or monitor closely the IPCC. However, I think it is an important step and fully support the initiative.

Can you describe a particular area of climate change that you are concentrating upon with research or a grassroots development project?

My main concern in dealing with climate change is risk management. I believe that despite all the on-going efforts, we need to be prepared to cope with the uncertainty associated with climate change. At the same time, I am promoting green and sustainable agricultural practice. In the long run, I believe that we have to go back to basics or adopt a more environmentally friendly livelihood and society.

Interview

Niall O'Connor is the Southeast Asia director of the Stockholm Environment Institute (SEI) in Bangkok, Thailand. The SEI is the leading environmental non-governmental organization (NGO) in the world. It awards the annual Stockholm Water Prize, known as the Nobel Prize for Water Rights and Development.

Can you describe the work of the SEI on climate change?

The SEI is an independent international research institute that has been engaged in environment and development issues at local, national, regional and global policy levels for more than a quarter of a century. The institute was formally established in 1989 by the Swedish government. Since then the SEI has established a reputation for rigorous and objective scientific analysis in the fields of environment and development. Our goal is to bring about change for sustainable development by bridging science and policy. We do this by providing integrated analysis to support decision makers.

The SEI's climate change work covers many areas in mitigation and adaption.

Adaption to climate change is one of the SEI's key areas of expertise— and one that cuts across a broad swathe of our work, involving specialists in everything from international climate policy and finance, to land- and water-resources planning, to local governance and capacity building.

For a number of years, the SEI has been one of the main actors in the development of methods and tools for assessing climate change vulnerability and impacts, for planning adaption and for sharing relevant knowledge. Another aspect of SEI's work addresses community-based and local adaptation, to understand the drivers of adaptive capacity, including how national and global development processes strengthen or weaken adaptive capacity at a local level, and increase or reduce vulnerability to climate risks.

The SEI previously implemented the Regional Climate Adaptation Knowledge Platform and currently undertakes an assessment of drought in the Mekong river region. The SEI also continues to support the development of weADAPT, a global adaption knowledge portal.

In the mitigation area, the SEI is involved in the development of an important energy planning tool, the Long-range Energy Alternatives Planning (LEAP) and a number of events with important implications for the global mitigation agenda. At Rio+20, SEI and its partners from around the world launched the results of a global assessment on sustainable energy for all. Energy access for household lighting, heating and cooking is widely regarded as a prerequisite to successfully fighting poverty and improving human well-being. Governments around the world increasingly recognize this, and the UN General Assembly named 2012 as the year of "Sustainable Energy for All."

The SEI has issued a report, "Energy for a Shared Development Agenda: Global Scenarios and Governance Implications," which explores the viability of such a goal through a global assessment of energy scenarios through 2050, case studies of energy access and low-carbon efforts around the world, and a review of the technological shifts, investments, policies and governance structures needed to bring energy to all. The goal of this and related SEI research is to lay out a more sustainable pathway for development in the global north and south alike, and encourage strategic discussions about how to achieve these ambitious but very important targets. World leaders can lay the groundwork at Rio+20, and they can increase the chances of success by making sustainable energy for all a priority in global climate and sustainable development policy.

Currently, the SEI has two major initiatives dealing with mitigation: Low Emission Development Pathways and Fossil Fuel Development and Climate Change Mitigation.

What concerns the SEI the most about climate change in your country.

Although I am Irish, I am currently based in Bangkok running the SEI's presence in Asia through its Asia center based here. It, therefore, behoves me to speak about the climate change realities from the region, Southeast Asia, where we currently operate. The impact of climate change in the region is particularly severe as it interacts with poverty. Although there are economic bright spots in the region, a significant number of countries here are still saddled by serious issues of poverty. Climate change will lead to deterioration in livelihoods and hence income, especially for those who are dependent on agriculture—a major source of employment and income for much of the rural sector in the region. Some areas in the region will suffer from the catastrophic impacts of sea level rise, which will drive increasing salinization. Together with land subsidence and increasing groundwater withdrawal, increasing salinity will also render farmlands unproductive, as seen now in the Mekong Delta. Sea level rise will also affect a significant portion of the region's inhabitants as estimates show that 70% of the region's population is located along coastal areas. This situation is particularly severe among countries with long coastlines and numerous islands such as Indonesia, the Philippines and Vietnam. Increasing temperatures will also lead to decreasing moisture and eventually drought in a number of countries in the region, notably northern Thailand and the dry zone of Myanmar. The impact of drought has caused households to suffer from declining agriculture income, food insecurity and increasing debt.

What concerns you most about climate change globally?

The SEI works across international and local levels. In doing so, we aim to ensure that policy debates are in harmony with local realities and to match local needs with larger-scale funding processes. The SEI interacts with a wide range of stakeholders and builds teams with diverse sets of skills, cultural backgrounds and perspectives, with experience in academia, policy and the private sector. From such a vantage point, we need to bring discussions about global climate change to grassroots level, where it matters most to the people who will be affected. No doubt those discussions are crucial and important, but if they do not include the harsh realities on the ground they remain noise amidst real problems of crippling poverty. So, to me, beyond the global bluster, we need to come to grips with how global climate change discourse is informed with the priorities and voices of the people who suffer most.

Can you describe a specific sustainability project in your area or country that is contributing to a "greener world" and lower carbon footprint?

We have a number of projects in the region but one which I take most pride in is the Sustainable Mekong Research Network (SUMERNET), a long-term program supported by the government of Sweden. Since its inception, it has addressed a number of climate change issues. Recently, it has facilitated an assessment of water scarcity in the Mekong region of which drought is an important manifestation. It has undertaken case studies in four countries using the robust decision support framework. Such studies will help governments in the region understand the threats of drought and develop appropriate policy interventions that will mitigate the causes and address the needs of vulnerable communities.

What steps could governments take to mitigate the growing threat from climate change?

The coming into force of the Paris Agreement in such a short period of time coupled with commitments from Parties to the Convention signals to me that the global community is now fully aware of the problem and is ready to walk the talk and get their hands dirty with real action. Still, what is missing, is how to get governments to wholeheartedly help poor households to anticipate, absorb and adapt to the impact of climate change. Curbing GHG emissions takes time; but while we wait, the impact of climate change is being felt now, especially among vulnerable communities. It is, therefore, critical that we not only talk but act. Acting now by developing projects that will build the resilience of poor and at-risk communities to climate change impact is an important order of business. As they say, think globally act locally. That said, I realize that adaption might only bring us so far. Analysts suggest that even if we maintain global temperature rise at less than 1.5 °C some loss and damage are still to be expected. Thus, I would like to see governments, especially in rich developed countries, not only provide funds for adaption but also help to address residual risks.

What is your reaction to the UNIPCC's accepted calculation by Prof. Myers of a minimum 200 million climate change refugees by 2050 and the world's readiness to cope with this humanitarian and environmental challenge?

I am glad that the issue of migration is raised, but I am careful about the language that is used. For one, migration is not necessarily a result of the impact of climate change. People migrate to enable their adaption as well as those who are left behind. In other words, migration is not only an outcome but also a means to sustain livelihoods and households. Prof Myers' message, while well-intentioned, may demonize processes of mobility and migration and fail to see it as an enabler of adaptation. Globally, we see people looking for better opportunities—that is, migration by people who did not suffer from persecution or environmental harm but were driven by exciting and profitable challenges. For another, the use of the word "refugee" is problematic. Such a word should not just be used in this way as it is a politically loaded term that is meant to describe people who are politically persecuted, whose lives are in danger and, hence, need to move. Such use of the term is highly emotive but will not serve any purpose except to possibly deny those who suffer political persecution a chance of claiming the term for themselves.

Can you describe a particular area of climate change that you are concentrating upon with research or a grassroots development project?

An area of climate change that we are currently passionate about is understanding how we, as an environmental think tank, can help countries in the region achieve their NDCs and contribute to their sustainable development goals. The implementation of the NDCs are opportunities for collaboration between governments and knowledge producers. We can encourage them to be ambitious in their commitments while reminding them of the need for transparency of the process. We can provide ideas. We can also inspire them through joint actions.

Our key priority at the moment is to help shape a policy agenda informed by evidence on how to build or enhance the capacities of poor households to anticipate, absorb and adapt to the impact of climate change. We want to develop a nuanced understanding of adaptive capacities and livelihood strategies amidst changing climate and extreme weather events. We also want to be able to develop approaches that will help households respond to residual risks. Here we intend to assess how far risk transfer mechanisms, such as insurance, catastrophe financing and risk pooling, can be vehicles for providing safety nets for vulnerable households while, at the same time, understanding how we can improve existing social protection mechanisms that have sustained these households for years, some even with minimal government presence. We believe that only in looking at the many dimensions of vulnerability and offering a comprehensive solution to the many risks that attend households in the region can we ensure that they are able to ride the impact of climate change with minimal suffering.

Interview

Prof. Shew-Jiuan Su is chairperson of the Department of Geography, National Taiwan Normal University, Taiwan. Trained as a human geographer, Prof. Shew-Jiuan Su focuses on the political economy and political ecology of hazardous areas. With increasingly extreme natural phenomena, she incorporates resilience in her research and compares the resilient strategies of areas with different geographies.

Can you describe your work related to climate change?

As a human and development geographer, I investigate how communities cope with environmental challenges and explore methods of adaption that are locally appropriate. The problems brought by global climate change are faced by and challenge human society. The best way of addressing these problems is to educate societies and communities in understanding how life style and daily activities can become adaptive and how to live resiliently in various geographies.

What concerns you the most about climate change in your country?

Extreme weather conditions are the greatest concern for Taiwan, in particular torrential rains and erratic typhoons. Located at the colliding edge of the Eurasian Plate and the Philippine Sea Plate, and at latitudes lying between tropical and subtropical zones, the island state of Taiwan has a fractured geology prone to many types of weather and erosion. With severe weather and erosion, hazards such as landslides and debris flows are easily triggered when there are torrential rains or typhoons, particularly after earthquakes strike. In addition, having a long coastline and most settlements being coastal, flooding and land subsidence usually become detrimental hazards.

What concerns you the most about climate change globally?

Food safety, food security and climate refugees are issues that I consider the most serious in the age of global climate change. Experiencing an erratic climate, where precipitation and temperature are not reliable or constant, society may have difficulty coping with changes that interfere with agricultural production, food safety and water contamination, to name just a few. As a consequence, people with fewer means will have more difficulty to cope with daily survival. Social instability will perhaps be part of the result. It in turn may bring political uncertainty that will partly lead to climate change refugees.

Can you describe a specific sustainability project in your area or country that is contributing to a "greener world" and a lower carbon footprint?

The Environmental Protection Administration of Taiwan has a carbon footprint reduction program entitled Low Carbon Sustainable Homeland, which is implemented throughout schools, institutes, businesses, communities and cities. In the program, 22 sustainable action plans are being conducted, including a low carbon community, green transportation, a culturally creative economy, among others. While reducing, recycling and reusing are typical means for a greener world, all actions taken in Taiwan are mostly enriched with a local and socio-cultural touch. The EPA in 2008 even affirmed a "No Regret Declaration of Conservation to Reducing Carbon Footprint," which engages daily activities, such as conserving water, using your own chopsticks instead of disposable ones, consuming local produce, eating more vegetables than meat, and so on, with the aim of reducing 1 kg of carbon emission per person per day.

What major new steps could your government take to mitigate the growing threats from climate change?

Policymaking to encourage factories to upgrade facilities and technologies is one important step that Taiwan's government can take to reduce carbon emissions. Taiwan's industrial structure contains many small- and medium-sized factories. Production costs are carefully calculated by such factories. Without policy promotion or political pressure, the cost of upgrading facilities will discourage factories from making changes. In addition, societal and political costs will be incurred, because implementing such policies needs to be sensitive to land use and local politics. Therefore, problems with industrial land use zoning will also have to be faced and resolved simultaneously to see the success of such policies.

What is your reaction to the UNIPCC's accepted calculation by Prof. Myers of a minimum 200 million climate change refugees by 2050 and the world's readiness to cope with this humanitarian and environmental challenge?

The UNIPCC's acceptance of Prof. Myers' estimate of climate migrants is laudable; it is better than being unconcerned. It is one way for the UNIPCC to raise awareness of global climate change. However, as a leading international scientific body, the UNIPCC could have been more cautious and knowledgeable—climate change hazards are not sufficient motivation to migrate. People facing hazards or problems do not necessarily relocate to solve those issues. Relocation also involves socio-cultural conditions. Within-boundary and trans-boundary migratory decisions are not easily made at any level. Due to socio-cultural and ethnic characteristics that influence people's identity and attachment to a location, migrating to a different place means and requires adaption. By and large, the world is not ready to cope with global environmental challenges, not even through receiving climate change refugees.

Can you describe a particular area of climate change that you are concentrating upon with research or a grassroots development project?

Because I was involved in the geo-park development of Taiwan, I designated a research program of working with geo-park communities to co-design plans for reducing carbon footprint. Specific plans and methods are yet to be finalized by consulting with local communities to make it a bottom-up approach. A general plan has been formed set, with three aspects. First, geo-park interpretation programs will be designed to guide visitors to walking through, instead of driving through geo-park sites.

The second is to develop food festival events for geo-park communities to serve visitors with local produce. A third aspect will focus on reusing local waste and producing handicrafts to enhance the local economy.

Interview

Dr. Lin is director of the Global Change Research Center at the National Taiwan Normal University in Taipei. He is a physical geographer whose research shows that the climate change issue is important to island ecology. "I have tried to let the global change research center act as a platform for this country. We have to start to encourage our young generation to pay more attention to this issue."

Can you describe your work related to climate change?

I am the director of the Global Change Research Center. Our center acts as a platform for government, academic and other related societies.

What concerns you the most about climate change in your country?

Natural hazards and extreme natural processes.

What concerns you the most about climate change globally?

Water resources and air pollution issues.

Can you describe a specific sustainability project in your area or country that is contributing to a "greener world" and lower carbon footprint?

Our local governments are encouraged to promote the use of substitute energy and have a low carbon consumption goal.

What major new steps could your government take to mitigate the growing threat of climate change.

Regional adaption policies have been proposed since 2015.

What is your reaction to the UNIPCC's accepted calculation by Prof. Myers of a minimum 200 million climate change refugees by 2050 and the world's readiness to cope with this humanitarian and environmental challenge?

There are regional differences. It is important to test and demonstrate the differences.

Can you describe a particular area of climate change that you are concentrating upon with research or a grassroots development project?

Water resources and energy issues are very important to Taiwan as it is a small island and quite vulnerable to climate change.

Notes

1. IPCC (2013). Climate Change 2013: The Physical Science Basis. Contribution of Working Group I to the Fifth Assessment Report of the Intergovernmental Panel on Climate Change. T.F. Stocker, D. Qin, G.-K. Plattner, M.M.B. Tignor, S.K. Allen, et al. (eds.). Cambridge University Press, Cambridge, UK, and New York. http://www.ipcc.ch/report/ar5/wg1/. Hijioka, Y., Lin, E., Pereira, J.J., Corlett, R.T., Cui, X., Insarov, G., Lasco, R., Lindgren, E. and Surjan, A. (2014). Asia. In Climate Change 2014: Impacts, Adaptation, and Vulnerability. Part B: Regional Aspects. Contribution of Working Group II to the Fifth Assessment Report of the Intergovernmental Panel on Climate Change. V.R. Barros, C.B. Field, D.J. Dokken, M.D. Mastrandrea, K.J. Mach, et al. (eds.). Cambridge University Press, Cambridge, UK, and New York. 1327–1370. https://www.ipcc.ch/report/ar5/ wg2/.
2. Dasgupta, P., Morton, J.F., Dodman, D., Karapinar, B., Meza, F., Rivera-Ferre, M. G., Sarr, T. and Vincent, K.E. (2014). Rural areas. In Climate Change 2014: Impacts, Adaptation, and Vulnerability. Part A: Global and Sectoral Aspects. Contribution of Working Group II to the Fifth Assessment Report of the Intergovernmental Panel on Climate Change. C.B. Field, V.R. Barros, D.J. Dokken, K.J. Mach, M.D. Mastrandrea, et al. (eds.). Cambridge University Press, Cambridge, UK, and New York. 1327–1370. https://www.ipcc.ch/report/ar5/wg2/.
3. Zhai, F. and Zhuang, J. (2009). Agricultural Impact of Climate Change: A General Equilibrium Analysis with Special Reference to Southeast Asia. ADBI Working Paper Series No 131. Asian Development Bank, Manila. http://www.adb.org/publications/agricultural-impact-climate-change-general-equilibrium-analysisspecial-reference.
4. Hijioka, Y., E. Lin, J.J. Pereira, R.T. Corlett, X. Cui, G.E. Insarov, R.D. Lasco, E. Lindgren and A. Surjan, 2014: Asia. In: Climate Change 2014: Impacts, Adaptation, and Vulnerability. Part B: Regional Aspects. Contribution of Working Group II to the Fifth Assessment Report of the Intergovernmental Panel on Climate Change, Barros, V.R., C.B. Field, D.J. Dokken, M.D. Mastrandrea, K.J. Mach, T.E. Bilir, M. Chatterjee, K.L. Ebi, Y.O. Estrada, R.C. Genova, B. Girma, E.S. Kissel, A.N. Levy, S. MacCracken, P.R. Mastrandrea, and L.L. White (eds.)]. Cambridge University Press, Cambridge, United Kingdom and New York, NY, USA, pp. 1327–1370.
5. TRF (2011).

6. Limsakul, A., A. Chidthaisong and K. Boonprakob (2011): Thailand's First Assessment Report on Climate Change 2011 (Working group I: Scientific Basis of Climate Change). THAI-GLOB, Thailand Research Fund.
7. Ibid.
8. Thammasart University Research and Consultancy Institute (2009): Final Report: Analysis of Sea Level Rise Impact to Land Use of Coastal Areas in Thailand.
9. Naeije, M., W. Simons, I. Trisirisatayawong, C. Satirapod and S. Niemnil (2012): Sea Level Rise and Subsidence in the Delta Area of the Gulf of Thailand. http://bit.ly/1TAEc9B.
10. Southeast Asia START Regional Center and World Wildlife Fund, 2008.
11. EIC Analysis/Note, "Thailand's drought crisis 2016: Understanding it without the panic" Teerayut Thaiturapaisan, 24 March 2016.
12. "Climate Risks and Adaptation in Asian Coastal Megacities" A Synthesis Report, World Bank, 2010.
13. Richard Anthes, Robert W. Corell, Greg Holland, James W. Hurrell, Michael C. MacCracken, and Kevin Trenberth, "Hurricanes and Global Warming: Potential Linkages and Consequences", American Meteorology Society 87, no. 5 (May 2006): 623–628.
14. Susmita Dasgupta, Benoit Laplante, Siobhan Murray, and David Wheeler, Sea-Level Rise and Storm Surges: A Comparative Analysis of Impacts in Developing Countries, Bangkok: World Bank, April 2009.
15. Patcharapim Sethaputra, National Capacity Self-assessment for Thailand: Final Report for the Convention of Climate Change (Bangkok: Office of Natural Resources and Environment, Ministry of Natural Resources and Environment of the Kingdom of Thailand, September 2009), http://chm-thai.onep.go.th/chm/pr/doc/CCC_Draft_Final_Report.pdf.
16. J. Alcamo, N. Dronin, M. Endejan, G. Golubev, and A. Kirilenko, "A new assessment of climate change impacts on food production shortfalls and water availability in Russia," Global Environmental Change, vol. 17, no. 3–4, pp. 429–444, 2007.
17. R. P. Allan and B. J. Soden, "Atmospheric warming and the amplification of precipitation extremes," Science, vol. 321, no. 5895, pp. 1481–1484, 2008.
18. Climate and Health Country Profile—2015 THAILAND, World Health Organization (WHO) and the U.N. Framework Convention on Climate, (UNFCCC) 2015.

19. MOAC (2012): Climate Change Strategy in Agriculture 2013–2016.
20. *Supnithadnaporn, Anupit; Inthisang, Jirapa; Prasertsak, Praphan; Meerod, Watcharin.* "Adaptation to Climate Change and Agricultural Sector in Thailand" (PDF). Asian Development Bank Institute (ADBI). Asian Development Bank.
21. Kisner, Corinne (July 2008). "Climate Change in Thailand: Impacts and Adaptation Strategies". Climate Institute.
22. *Wangkiat, Paritta (27 November 2016).* "The heat is on". *Bangkok Post.*
23. Myanmar's National Adaptation Programme of Action (NAPA) to Climate Change, Government of Myanmar, 2012.
24. Ibid.
25. Swe, K.L. Review of adaptations in socio-economical sectors in Myanmar. Yezin Agricultural University.
26. Tripartite Core Group. Post Nargis Joint Assessment (PONJA) "Post Nargis Periodic Review II". 2009.
27. Union of Myanmar et al. 2009. Hazard Profile of Myanmar.
28. Swe, K.L. Review of adaptations in socio-economical sectors in Myanmar. Yezin Agricultural University.
29. Kovats, S. and Akhtar, R. 2008. Climate, climate change and human health in Asian cities. Environment and Urbanization 20: 165–175.
30. Ibid.
31. Prof. Dr. Soe Lwin Nyein. 2010. Climate change on communicable diseases. Director of Epidemiology, Ministry of Health.
32. Kovats, S. and Akhtar, R. 2008. Climate, climate change and human health in Asian cities. Environment and Urbanization 20: 165–175.
33. Myanmar's National Adaptation Programme of Action (NAPA) to Climate Change, Government of Myanmar, 2012.
34. Ibid.
35. EJF (2009) "No Place Like Home—Where next for climate refugees?" Environmental Justice Foundation: London.
36. Cazenave A., Llovel W. (2010) Contemporary sea level rise. Annu. Rev. Mar. Sci. 2 (1), 145–173. https://doi.org/10.1146/annurev-marine-120308-081105.
37. Suzuki T. (2009) Estimation of inundation damage caused by global warming in three major bays and western parts of Japan. Global Environment, Association of International Research Initiatives for Environmental Studies 14 (2), 231–236.
38. Ministry of Land, Infrastructure, Transport and Tourism.
39. Oouchi K., Yoshimura J., Yoshimura H., Mizuta R., Kusunoki S., Noda A. (2006) Tropical cyclone climatology in a global warming climate as

simulated in a 20 km-mesh global atmospheric model: frequency and wind intensity analyses. J. Meteorol. Soc. Jpn. 84, 259–276. https://doi.org/10.2151/JMSJ.84.259.
40. Kitoh A., Ose T., Kurihara K., Kusunoki S., Sugi M., KAKUSHIN Team-3 Modeling Group (2009) Projection of changes in future weather extremes using super-high-resolution global and regional atmospheric models in the KAKUSHIN Program: Results of preliminary experiments. Hydrological Research Letters 3, 49–53. https://doi.org/10.3178/HRL.3.49.
41. Mimura, Nobuo. "Sea-Level Rise Caused by Climate Change and Its Implications for Society." Ed. Kiyoshi Horikawa. Proceedings of the Japan Academy. Series B, Physical and Biological Sciences 89.7 (2013): 281–301. PMC.
42. "Japan to Suffer Huge Climate Costs" Alva Lim, United Nations University, Brendan FD Barrett, Royal Melbourne Institute of Technology, June 30, 2009.
43. "Consolidated Report on Observations, Projections and Impact Assessments of Climate Change: Climate Change and Its Impacts in Japan" Ministry of the Environment, 2012.
44. Prof. Nobuo Mimura, Director of the Institute for Global Change Adaptation Science at Ibaraki University, the "Global Warming: Impact on Japan" Japan Ministry of Environment.
45. Tomoya Watanabe (2012) "What level does climate change make impacts on agricultural production, how to address and how should agriculture, forestry and fishery sector tackle climate change? Research project funded by the Ministry of Agriculture, Forestry and Fisheries.
46. "Consolidated Report on Observations, Projections and Impact Assessments of Climate Change: Climate Change and Its Impacts in Japan" Ministry of the Environment, 2012.
47. Augendre Marie. "Risques et catastrophes volcaniques au Japon: enseignements pour la géographie des risques" in Habiter les territoires à risques. Presses polytechniques et universitaires romandes. 185–207. 2012.
48. Isobe M. (2013) Impact of global warming on coastal structures in shallow water. Ocean Engineering in press.
49. Mimura, N. "Sea-Level Rise Caused by Climate Change and Its Implications for Society." Ed. Kiyoshi Horikawa. *Proceedings of the Japan Academy. Series B, Physical and Biological Sciences* 89.7 (2013): 281–301. *PMC*. Web. 24 Apr. 2017.

5

China

"The best time to plant a tree was 20 years ago."
Chinese Proverb

Introduction

China is one of the nations in the world most profoundly affected by climate change. It is impacted at both ends of the environmental spectrum with current and exponentially increasing drought in the north and flooding in the south. The challenges are enormous and will consume considerable human, economic and social costs for the nation of 1.4 billion citizens. The land area is 9,706,961 km^2 and the coastline measures 14,500 km. Agricultural lands cover approximately 54% of the total with 11% devoted to permanent crops and 41% to pasture lands. Agriculture is an industry that will be significantly affected by climate variation. The most common environmental threats include flooding, drought, desertification, saltwater intrusion, typhoons, land subsidence, earthquakes and tsunamis. The GDP is 11.3 trillion and the country has an HDI rank of 90. China is one of the world's largest CO_2 emitters at 28.2% (2017) of the global total.

Flooding and Drought

Severe flooding in southern China has been a constant for many centuries. This is a climate trend that is anticipated to increase in intensity over the twenty-first century. In 1996, official national statistics showed 200 million people affected by flooding: there were more than 3000 deaths, 363,800 people were injured, 3.7 million houses were destroyed and 18 million houses were damaged. Direct economic loses exceeded USD 12 billion.[1] In 1998, official national statistics showed 200 million people affected by flooding, more than 3000 deaths and 4 million houses damaged; direct economic losses exceeded USD 20 billion.[2]

China represents a climate dichotomy in many respects with severe drought in the north that is expected to intensify in the twenty-first century and devastating flooding in the south. The metropolis of Shanghai, one of the most populated urban centers in the world, is already experiencing the devastating impacts of coastal flooding and saltwater intrusion. "Based on the SRES emission scenario A1B for atmospheric greenhouse gasses and aerosols, a number of climate models and the RegCM3 simulations have been performed to project the drought and flood risk changes over the twenty-first century. These simulations consistently indicate that most parts of China will experience more frequent, more intense, and much longer flood events under a global warming scenario. The simulation from the RegCM3 indicates that the proportion of the flood area in China increases from 11.2% in the period of 1980–1999 to 25.4% in the period of 2080–2099."[3]

Recent research at Oxford University's Environmental Change Institute has identified cities in China that are the most vulnerable to infrastructure damage due to climate change. The critical areas include Beijing, Tianjin, Jiangsu, Shanghai and Zhejiang. It is well documented that China will experience increasing climate change effects such as flooding, desertification, saltwater intrusion, drought, severe weather episodes, extreme heat waves, mudslides and wind storms. Infrastructure damage due to severe repercussions of climate change will absorb billions of dollars in mitigation and repair costs. "In the case of China, the impacts on businesses are potentially astronomical given the scale of its manufacturing production (30% of its GDP in 2010) and its role in the global supply chain. Already in some of the more economically developed coastal provinces, flooding costs, for instance, at city and industrial levels, account for more than 60% of the total flooding costs. The 2011 floods alone, for example, resulted in the interruption of services to 28 rail links, 21,961 roads and 49 airports, and the failure of 8516 electricity transmission lines."[4] Moreover, there will be substantial burdens upon human health and development that are incalculable.

In 2016, farmers lost an estimated USD 1.2 billion from drought. With increasing water insecurity and drought conditions in northern China becoming more prevalent, the conditions are ripe for severe long-term impacts. Already, Chinese industry and agriculture consume about 85% of the country's fresh water resources. Complicating matters further is the fact that China has 20% of global population yet only 7% of global fresh water. In an effort to mitigate the extreme water shortage crisis facing northern China, the government launched an ambitious irrigation and water diversion project for the north. To date, USD 60 billion has been spent on the project by the government, which includes 2700 miles of constructed water systems designed to provide water security in the driest regions of the country.

In 2014, the Chinese province of Shandong suffered the worst drought it had experienced in half a century, leaving reservoir beds chapped and the peanut crop (which is typically drought resistant) withering on the vine. This dry spell wasn't just one more offshoot of global warming; instead, it was the result of booming industrial activity that had been sapping Shandong's notoriously thin water supply. A recent study revealed that average annual water availability per capita in Shandong is only 11,300 cubic feet (320 m^3), less than one-sixth of China's average or about one-twenty-sixth of the global average. In 2011, large areas of northern China experienced the worst drought in 60 years with some areas going 5–6 months without adequate rainfall. This situation resulted in damage to 5 million ha (12.4 million acres) of cropland. That same year the Chinese government provided CNY 2.2 billion (USD 334 million) in drought relief to stricken farmers. In 2010, drought affected large areas of Guizhou, Yunnan and Sichuan provinces with rainfall of at least 60% below normal levels. In the province of Guizhou, an astonishing 86 out of its 88 cities fell within the designated drought critical zone.

Extreme Weather

According to the IPCC, climate change is likely to increase the frequency and intensity of weather events that are sufficiently extreme to have a major impact on China's economy, environment and society.[5] "In 2001 a drought caused temporary shortages in drinking water for 33 million rural people and 22 million livestock and cost China an estimated 6.4 billion United States dollars (US$) in lost crop production."[6]

Another major challenge facing China that is exacerbated by climate change is desertification, a process whereby land becomes parched and untenable for agricultural production and there is massive loss of vegetation cover. Deforestation is a major contributing factor. In 2018, approximately 2.6 million km^2 or 25% of the land mass across 18 provinces is affected by desertification. Due to serious concerns over desertification in northern China, the government initiated a large-scale reforestation program in 1978 that involves the planting of an estimated 66 billion trees with a completion target date of 2050. The project has been called the "Green Wall of China." Sandstorms are symptomatic of desertification and have devastating effects upon people and land. In 1999, 85 people died in China's worst sandstorm for loss of human life and a destructive storm in 2010 affected 270 million people in 16 provinces. Desertification threatens 400 million people in China, mostly in western and northern China and creates a financial burden of CNY 45 billion annually.

According to one report, for seriously desertified regions, the loss amounts to as much as 23.16% of annual GDP. The fact that one-third of the country's land area is eroded has led some 400 million people to struggle to cope with a lack of productive soil, destabilized climatological conditions and severe water shortages. Droughts damage about 160,000 km^2 of cropland each year, double the area damaged in the 1950s.[7]

Health, Human Development and Food Security

The impacts of climate change upon human health are pervasive and include air pollution, extreme heat, flooding, the rise of infectious diseases, drought, infrastructure damage, climate induced migration, contaminated water sources and inadequate food systems that have been damaged or degraded. As the leading coal using nation in the world, the effect on air quality is negative. China now ranks air pollution as the leading cause of death. "The short-term, direct impact of climate change on health in China is likely to be via increased air temperatures and increased air pollution, which have already substantially increased morbidity and mortality."[8] Aside from declining air quality in urban China, there are multiple health risks facing the population. Flooding is also a major health and human development risk in the country. According to Debarati Guha-Sapir, a director of the WHO Collaborating Centre for Research on the Epidemiology of Disasters (CRED), "Urban flooding and rural flooding can have different impacts on the populations affected. In rural areas, there is definitely a very high risk of diarrhoea and diseases because of contamination of water. You can also have an increase in breeding sites of mosquitoes or other vectors, and in China in particular both malaria and dengue are an increasing problem, as is schistosomiasis, a disease carried by snails. The strain of schistosomiasis that you see in China was a very, very big problem and then they got rid of all the snails—they mobilised millions of people. Now it's coming back again. These are all diseases of concern and do have a link to the heavy precipitation and increases in breeding sites."[9]

One of the severe consequences of climate change is the well documented effect on food systems and productive agricultural lands. China will have a projected population of 1.41 billion citizens by 2050, yet will be confronted with extreme weather and climate conditions that will clearly jeopardize food security. Climate scientists have already predicted that food prices will skyrocket in many countries due to climate change and will place disproportionate burdens upon impoverished citizens and developing countries. An estimated 80 million Chinese are living below the poverty line and there are 100 million migrants in the country with no stable employment or living accommodation. Among the groups most impacted by climate change are the poor who represent a demographic that is often malnourished and whose food purchases are severely affected by rising commodity prices. The IPCC reports that food prices, particularly for wheat, could rise by between 3 and 84% by 2050.

Currently, approximately 50% of agricultural land is utilized for the cultivation of maize, rice and wheat. The Fifth Assessment Report of the Intergovernmental Panel on Climate Change indicated that climate change is already having a negative impact on agriculture and food production by adversely affecting major crops, livestock production and fisheries.[10] "It remains unclear whether China can feed its entire population adequately in the twenty-first century."[11]

In China, 100 million farmers and millions of farm migrants are severely threatened by agricultural land destruction and low crop yields from varying climate effects such as extreme heat, desertification, flooding and sandstorms. Climate change is likely to exacerbate the problems faced by these people, who often lack the financial and other resources needed to respond effectively.[12] Unless carbon dioxide fertilization provides unexpected benefits, the overall effects of agricultural land conversion, climate change and reduced water availability could reduce China's per-capita cereal production, compared with that recorded in 2000, by 18% by the 2040s.[13] By 2030–2050, loss of cropland resulting from further urbanization and soil degradation could lead to a 13–18% decrease in China's food production capacity—compared with that recorded in 2005.[14]

An analysis of projected climate change phenomena in China in the twenty-first century concluded with somber scenarios that will severely effect health, economics and development for generations to come. "The average temperature increments at the end of the twenty-first century over China would be over 3% while the percentage of precipitation would increase by 10%, an overall increase of the maximum/ minimum surface air temperature during the 2080s, the occurrence frequency of extremely high temperature and extreme precipitation events would increase, low temperature events would decrease; drought with high temperature events in the northern part of China, the flooding in summer and drought in winter in southern part of China would be enhanced with transient atmospheric GHGs concentration increasing in future."[15] A further study based upon precipitation patterns and future trends noted that, "The results revealed that severe and extreme droughts have become more serious since late 1990s for all of China, and persistent multi-year severe droughts were more frequent in North China, Northeast China, and western Northwest China."[16]

Although extreme weather events are not new to China, the severity and frequency are phenomena that are of increasing concern to Chinese authorities. While it is true that China is a rapidly developing country and records some of the most impressive annual economic growth figures of any nation, there are still more than 100 million Chinese citizens who live in conditions of poverty and thus are unable to adequately cope economically or medically with the devastation and dislocation of climate change. Moreover, the economic costs of climate change are a major impediment to national and regional development. "From 1960–2013, the country endured 784 simultaneous rainstorms that were concurrently recorded in over 10 stations, i.e., an average of 14.5 times a year. During this period, the number of simultaneous rainstorm events increased from 13.5 to 17.3 a year; two additional heat wave events occurred between 1997 and 2008 when compared with 1976–1994 in terms of annual average frequency."[17] Between 1984 and 2013, weather and climate disasters caused direct economic losses of CNY 188.8 billion per year, which is equivalent to 2.05% of the GDP.[18]

Climate Change Refugees

Although China is not expected to generate external climate refugee flows the there is a strong eventuality of future mass internal migration due to flooding, drought and extreme weather events. Since 2003, in just one region of the country, 480,000 people in inner Mongolia have been relocated in the on-going battle against desertification. One of the critical concerns for China is internal climate refugee flows from the increasing incidence of northern drought and southern flooding. Moreover, coastal flooding that will devastate cities such as Shanghai may generate millions of internally displaced people in the twenty-first century.

In 2007, a study used satellite data to analyze coastal areas in 224 countries with elevations of less than 30 feet above sea level. Notable conclusions from the research were that 634 million people inhabit these low-lying areas, which represents 10% of global population situated in only 2% of the world's land mass. Two-thirds of the world's largest cities—those with more than 5 million residents—are at least partially located there. "The study also observed that 10 countries with the largest share of their populations in low-elevation coastal zones are Bangladesh, China, Egypt, Gambia, India, Indonesia."[19] A significant number of Chinese citizens will inevitably be forced away from coastal areas due to growing coastal flooding and saltwater intrusion threats.

Another example is the dry and drought prone isolated region of Xihaigu that is generating an increasing number of environmental refugees. The region can sustainably accommodate 1.3 million people according to the Chinese Academy of Sciences and the Ministry of Land and Resources, yet had a population of 2.4 million in 2016. A relocation program was launched in 2011 that aimed for a total of 350,000 people slated for environmental resettlement.

The Tibetan Plateau

The ancient kingdom of Tibet was annexed by China in 1950. The Tibetan Plateau is known as one of the "Three Poles" in the world, the other two being the North Pole in the Arctic and South Pole in the Antarctica. The Tibetan Plateau, with an estimated 46,000 glaciers, is the source of all ten of the major rivers systems in South and Southeast Asia including the Mekong, Ganges, Brahmaputra, Indus, Sutlej, Irrawady and Salween, with a combined river dependent population of approximately 2 billion people. India, China, Nepal, Bhutan, Bangladesh, Pakistan, Vietnam, Myanmar, Cambodia, Laos and Thailand all have rivers that flow from the Tibetan Plateau.

A trend of alarming concern to policymakers and climate scientists and ultimately more than 2 billion citizens in several countries is the rapid rate of glacier melting. This phenomenon will inevitably lead to massive flooding of river systems that support lives and economic livelihoods for hundreds of millions of citizens in multiple countries. Moreover, the Chinese practice of damming sections of several rivers has had an impact on downstream communities. With water security becoming an ever more critical issue, this practice will exacerbate regional tensions—particularly Indo–Sino relations, which are already fragile since the 1962 war and the Chinese occupation of Aksa Chin in the Kashmir.

The construction of dams on the Mekong, Brahmaputra and Salwen rivers is cause for concern among regional neighbors. According to a recent report, "China has constructed seven dams along the Mekong River in Tibet and 21 more are planned. Almost 60 million people depend upon the river for food and water security in Cambodia, Laos, Thailand and Vietnam. Any alteration to its flow could have dire consequences for them, including the creation of environmental refugees. The Salween River is a World Heritage Site and home to 25 per cent of animal species in the world. China has constructed a dam 5.5 kilometres away from the Heritage Listed area, with plans for more dams in the pipeline. The construction of further dams would not only pose a risk to the ecological preservation of the Salween but could cause seawater intrusion in downstream Myanmar. China's Zangmu Dam became operational in October 2015 and is situated along the upper reaches of the Yarlung Tsangpo (known as the Brahmaputra in India). Due to its close proximity to India, the dam may trigger floods in the Indian state of Assam during the rainy season and may cause the Brahmaputra to dry up during winter."[20] With respect to India, 100 million people in India's Assam state are dependent upon the waters of the Brahmaputra river, thus damming of the river upstream by China will have major negative consequences for those citizens.

Numerous reports have cited global warming induced glacier melting of 7% annually on the Tibetan Plateau. The increased ice melt from the plateau will inevitably lead to severe flooding, soil erosion, crop destruction and economic devastation. "A combination of urbanization, intensified militarization linked to China's strategic aims, infrastructure construction and warming temperatures are creating an 'ecosystem shift' in Tibet. This involves irreversible environmental damage, including the predicted disappearance of large areas of grasslands, alpine meadows, wetlands and permafrost on the Tibetan plateau by 2050, with serious implications for environmental security in China and South Asia."[21]

There must be intensified efforts and policy steps from all nations sharing the vital waterways of the Tibetan Plateau to efficiently and ecologically manage environmental conditions in the face of ever increasing climate change and anthropogenic threats.

City Profile: Shanghai

Shanghai is a leading global commercial center and sea port with a population of 24 million. It is located on the eastern coast of the country. One of the serious climate change risks to the city is flooding that is exacerbated by an average sea level elevation of only 4 m. The general patterns of global warming and sea level rise are climate phenomena that are expected to intensify over the coming decades, thus placing millions of Shanghai citizens at risk. The city area covers 6340.5 km^2 and the water area is 122 km^2.

Climate change impacts fall hardest upon vulnerable groups such as the health challenged, physically impaired, children, impoverished and the elderly. In Shanghai, there were 4 million citizens over the age of 60 in 2016. According to a recent study, "Shanghai has become more vulnerable to high-impact weather and meteorological disaster, especially precipitation extreme, summer high temperature, haze and typhoon, so more strategies of mitigation and/or adaptation of natural disasters are quite useful and necessary for local government and the public in the future."[22]

A report by the US-based research group Climate Central documented the repercussions of global warming on the world's coastal areas and mega-cities. "It states that a 4 degree Celsius increase in the earth's weather will cause sea levels to rise enough to submerge coastal areas, leaving 470 to 760 million people's homes underwater." Major cities such as Shanghai, Hong Kong, London, New York, Sydney and Mumbai could end up submerged by 2100. If we could manage to cut carbon emissions so there was only a 2 °C increase, only 130 million people would be affected.[23] The UN has identified risk threats in low elevation coastal zone (LECZ) areas. Some of the most vulnerable Asian and South Asian cities include Shanghai, Bangkok, Tokyo, Mumbai, Kolkata, Manila and Jakarta. "It has been estimated that, in the absence of any other changes, a sea-level rise of 38 cm would increase five-fold the number of people affected by storm surges."[24] By 2050, projections reveal a sea level rise in the East China Sea of between 7.5 and 14.5 cm. Recent studies undertaken by the UK's Met Office show a "best estimate" of 4 °C being reached by 2070, with a possibility that it could come as early as 2060.[25] Thus the scenario for Shanghai of a 4 °C rise in temperature by the latter quarter of the twenty-first century will predictably be the cause of devastating health, economic and social costs that in turn will propel millions of Shanghai citizens from their homes and communities as displaced internal environmental refugees. Even the best case scenario of a 2 °C rise in temperature by the same time benchmark will activate harsh consequences. Shanghai clearly faces more urgent climate change mitigation demands due to flooding compared to any other population center in the country. Despite annual GDP growth rates of 6–10% from 1995 to 2017, as a developing country China will be developmentally imperiled by the massive human and financial burdens of climate change destruction.

Innovation and Adaption Strategies

China is the world leading emitter of GHG emissions. Since 1990, emissions have risen 80%, with heavy coal use being a determining factor.

September 3, 2016, China ratified the Paris Agreement.

China is on target to meet CO_2 emissions between 2025 and 2030, which is an essential component of the Nationally Determined Contribution (NDC) commitment under the Paris Agreement. "The targets set in the NDC, which include the target to peak CO_2 emissions by 2030 at the latest, lower the carbon intensity of GDP by 60%–65% below 2005 levels by 2030, increase the share of non-fossil energy carriers of the total primary energy supply to around 20% by that time, and increase its forest stock volume by 4.5 billion cubic meters, compared to 2005 levels. Total GHG emissions are likely to continue increasing until 2030, as China has not yet implemented sufficient policies addressing non-CO_2 GHG emissions (methane, nitrous oxide, HFCs etc.). This indicates a need for further action in this area, and it is encouraging that the NDC acknowledges that addressing these gases is important."[26]

- March 16, 2016, the Chinese Government officially approved its 13th five-year plan, which is a plan for economic and social development between 2016 and 2020. Included in the plan are reductions in water use, energy use and CO_2 emissions of 23%, 15%, and 18% respectively by 2020.
- With air pollution being the leading cause of fatalities in China, there is an important goal to have good air quality, defined as the density of fine particle matter, on at least 80% of the days in a year by 2020.
- President Xi Jinping set a national goal to cap CO_2 emissions by 2030 and to reduce the emission of CO_2 per unit of GDP by 60–65% from the 2005 level by 2030.
- June 4, 2007, China announced its first National Climate Change Plan.
- 1993 Ratification of UNFCC.

Interview

Zamlha Tempa Gyaltsen is a research fellow at the Tibet Policy Institute and also heads the Environment and Development Desk of the Tibet Policy Institute. The focus of his research is the environmental situation in Tibet and the social, environmental impacts of climate change on the Tibetan Plateau.

Can you describe your work related to climate change?

Our focus of research is on the current environmental situation inside Tibet. I examine the social and environmental impact of climate change, how the climate change induced temperature rise on the Tibetan Plateau impacts the social, economic and environmental state of the Tibetan Plateau and the Tibetan people and work out how best we could mitigate and adapt to the inevitable change and survive successfully as Tibetans have done for thousands of years on the world's highest plateau.

What concerns you the most about climate change in your country?

The biggest concern for the future of Tibet due to climate change is the rapid grassland desertification in many parts of the Tibetan Plateau. Climate change and rising temperatures have resulted in fast permafrost degradation leading to alarming grassland desertification in many parts of north and northeastern Tibet. Desertification could severely impact the social, economic and environmental state of Tibet and the Tibetan people. The increasing number of natural calamities (landslides, floods or mud floods and glacial lake outburst floods) in Tibet is another concern for the future. This year (2016) Tibet has experienced an unprecedented number of natural disasters, including glacial avalanches, floods, heavy rainfall and drought.

What concerns you the most about climate change globally?

The increasing frequency of natural disasters around the world is the biggest concern. The potential water crisis and food shortages in certain parts of the world could cause conflict and social unrest.

What steps could your government take to mitigate the growing threats from climate change?

A sincere, well researched, well organized and local participatory tree-plantation drive suitable to local environmental conditions could greatly mitigate the possible negative impact of climate change on the Tibetan Plateau.

Can you describe a particular area of climate change that you are concentrating upon with research or a grassroots development project.

The area that I am focusing on is the possible social environmental impact of climate change on the Tibetan Plateau.

Interview

Tingju Zhu is a research fellow at the International Food Policy Research Institute where he conducts interdisciplinary research at the interface of sustainable water management, food security and resource economics towards a water- and food-secure future, focusing on water issues in developing countries.

Can you describe your work related to climate change?

I conduct the assessment of climate change impact on water resources and agricultural production at various levels and identify adaption measures at both technical and policy levels.

What concerns you the most about climate change in your country?

I am most concerned about glacier retreat and loss of snowpack; increasing frequency and severity of droughts and floods; and crop yield decline under climate warming.

What concerns you the most about climate change globally?

Globally, I am concerned about increasing hydro-meteorological extreme events, food security and conflicts.

Can you describe a specific sustainability project in your area or country that is contributing to a "greener world" and lower carbon footprint?

Promoting the use of electric cars is one such project.

What steps could your government take to mitigate the growing threats from climate change?

The government could reform its energy policy.

Can you describe a particular area of climate change that you are concentrating upon with research or a grassroots development project?

By developing and applying systems analysis and integrated assessment models, I analyze climate change impact on water resources, such as irrigation water supply and flood protection, and on agricultural production and food security in a number of countries in Asia and Africa, as well as the state of California in the USA.

Notes

1. IFRC, 1997.
2. National Climate Centre of China, 1998.
3. Chen Huo-Po, Sun Jian-Qi & Chen Xiao-Li (2013), Future Changes of Drought and Flood Events in China under a Global Warming Scenario, Atmospheric and Oceanic Science Letters, 6:1, 8–13 To link to this article: https://doi.org/10.1080/16742834.2013.11447051.
4. Xi Hu, "Where will climate change impact China most?" World Economic Forum, April 5, 2016.

5. Intergovernmental Panel on Climate Change. Climate change 2014: impacts, adaptation, and vulnerability. Summary for policymakers. Working Group II contribution to the Fifth Assessment Report of the Intergovernmental Panel on Climate Change. Cambridge: Cambridge University Press; 2014. Available from: http://ipcc-wg2.gov/AR5/images/uploads/IPCC_WG2AR5_SPM_Approved.pdf.
6. Shen C., Wang W., Hao Z, Gong W. Exceptional drought events over eastern China during the last five centuries. Climate Change. 2007; 85 (3–4):453–471. https://doi.org/10.1007/s10584-007-9283-y.
7. "China is quickly turning into a desert, and it's causing problems across Asia" Marijn Nieuwenhuis, The Conversation UK, Business Insider, May 16, 2016.
8. Kan H., Chen R. Tong S. Ambient air pollution, climate change, and population health in China. Environ Int. 2012 Jul; 42:10–9. https://doi.org/10.1016/j.envint.2011.03.003. PMID:21,440,303.
9. "China faces a flooding crisis as natural disasters triple in 30 years" Olivia Boyd, China Dialogue, January, 1, 2013.
10. Intergovernmental Panel on Climate Change. Climate change 2014: impacts, adaptation, and vulnerability. Summary for policymakers. Working Group II contribution to the Fifth Assessment Report of the Intergovernmental Panel on Climate Change. Cambridge: Cambridge University Press; 2014. Available from: http://ipcc-wg2.gov/AR5/images/uploads/IPCC_WG2AR5_SPM_Approved.pdf.
11. Zhang Z., Duan Z., Chen Z., Xu P., Li G. Food security in China: the past, present and future. Plant Omics J. 2010; 3:183–9.; Ye L., Tang H., Wu W., Yang P., Nelson G., Mason-D'Croz D., et al. Chinese food security and climate change: agriculture futures. Economics. 2014; 8(2014–1):1–39. https://doi.org/10.5018/economics-ejournal.ja.2014-1; Huang J., Rozelle S. Agriculture, food security, and poverty in China. Past performance, future prospects, and implications for agricultural R&D policy. Washington: International Food Policy Research Institute; 2009. http://lib.icimod.org/record/14368/files/4068.PDF?version=1.
12. Knox J., Hess T., Daccache A., Wheeler T. Climate change impacts on crop productivity in Africa and South Asia. Environ Res Lett. 2012; 7(3):034032. https://doi.org/10.1088/1748-9326/7/3/034032.
13. Xiong W., Conway D., Lin E., Xu Y., Ju H., Jiang J., et al. Future cereal production in China: modelling the interaction of climate change, water availability and socioeconomic scenarios. Glob Environ

Change. 2009; 19(1):34–44. https://doi.org/10.1016/j.gloenvcha.2008.10.006, Publication: Bulletin of the World Health Organization; Type: Policy & practice Article ID: BLT.15.167031.
14. Ye L., van Ranst E. Production scenarios and the effect of soil degradation on long term food security in China. Global Environ. Change. 2009; 19(4):464–81. https://doi.org/10.1016/j.gloenvcha.2009.06.002).
15. "Statistical Analyses of Climate Change Scenarios over China in the 21st Century" Xu Yinlong, Huang Xiaoying, Zhang Yong, Lin Wantao, Lin Erda, Advances in Climate Change Research, Adv. Clim. Change Res., 2006, 2: 1673–1719 (2006).
16. "Are droughts becoming more frequent or severe in China based on the Standardized Precipitation Evapotranspiration Index: 1951–2010?" Meixiu Yu, Qiongfang Li, Michael J. Hayes, Richard R. Heim in Journal of Climatology, Volume 34, Issue 3, 15 March, 2014.
17. Yun Gao, "China's response to climate change issues after Paris Climate Change Conference" in Advances in Climate Change Research, Vol. 7, Issue 4, December 2016.
18. D.H. Qin, "China National Assessment Report on Risk Management and Adaptation of Climate Change" National Science Press, Beijing (2015).
19. "New Report and Maps: Rising Seas Threaten Land Home to Half a Billion" Climate Central, November 8, 2015.
20. "Tibet: A Major Source of Asia's Rivers" by Madeleine Lovelle, Global Food and Water Crises Research Programme, Future Directions International, February 4, 2016.
21. "New report reveals global significance of Tibet, earth's Third Pole, and challenges China's policies" International Campaign for Tibet, December 8, 2015.
22. Jun Shi and Linli Cui, "Characteristics of high impact weather and meteorological disaster in Shanghai, China" 2012.
23. "New Report and Maps: Rising Seas Threaten Land Home to Half a Billion" Climate Central, November 8, 2015.
24. EJF (2009) No Place Like Home—Where next for climate refugees? Environmental Justice Foundation: London.
25. Findings presented by UK Met Office Hadley Centre, 28/09/2009. http://www.metoffice.gov.uk/climatechange/news/latest/four-degrees.html.
26. Climate Tracker, http://climateactiontracker.org/countries/china.html, 2017.

6

Africa: Kenya, South Africa, Botswana

"Treat the earth well; it was not given to you by your parents; it was loaned to you by your children."
Kenyan Proverb

Introduction

The African continent and population, compared to all regions in the world, will experience the most difficult and dramatic consequences of climate change that will include widespread human suffering and environmental devastation. Severe drought; intense heat waves; constant coastal flooding episodes; elevated health crises; food insecurity that will be devastating in many regions; climate change induced political instability; declining agricultural production and hence employment dislocation; and massive climate refugee flows both externally and internally are scenarios that will emerge with certainty. According to the IPCC, relative to low socio-economic conditions, the impact of weather related disasters in poor countries may be 20–30 times larger than in industrialized countries.[1] Africa is the second largest continent and comprises approximately

20% of global land surface. It borders five major bodies of water, the Atlantic, the Pacific and Indian oceans and the Mediterranean and Red seas. Thus sea level rise, flooding and saltwater intrusion are massive threats to coastal populations and infrastructure.

Africa will experience high population growth rates in coastal regions that will coincide with sea level rise and flooding. The combination of these two events poses serious health and financial burdens for coastal countries and populations. A recent report on coastal flooding stated that, "In particular, the LECZ (low elevation coastal zone) population of Sub-Saharan Africa (all of Africa except Northern Africa; includes the Sudan), which represented 45% of the African nations' LECZ population in 2000, could grow from 24 million (2000) to 66 million by 2030 and to 174 million by 2060 due to an average coastal growth rate of up to 3.3% (2000–2030) and 3.2% (2030–2060). These rates are considerably higher than in Asia, where annual rates of growth are expected to reach 1.4% in the first three decades (2000–2030) and afterwards drop to 1.2%."[2] Moreover, "by 2050, between 350 million and 600 million people are projected to experience increased water stress due to climate change. Climate variability and change is projected to severely compromise agricultural production, including access to food, in many African countries and regions. In many African countries, existing health threats—such as malnutrition, malaria and other vector-borne diseases—can be exacerbated by climate change."[3] Climate change projections for the African continent include: 75–250 million people are projected to be exposed to increased water stress by 2020 due to climate change; yields from rain-fed agriculture could be reduced by up to 50%; agricultural production, including access to food, in many African countries is projected to be severely compromised; the cost of adaption to coastal countries could amount to at least 5–10% of their GDP; and by 2080, an increase of 5–8% in area of arid and semi-arid land is predicted.[4]

Kenya

Kenya has a population of 48 million (2017), is situated on the east coast of Africa along the Indian Ocean and has a land area of 582,646 km² (224,961 square miles) and a coastline of 536 km. Kenya borders

Ethiopia, Somalia, South Sudan, Tanzania and Uganda. Approximately 48% of the land is devoted to agriculture. Current environmental problems include flooding and episodic drought, which will be exacerbated by the effects of climate change. The GDP is USD 69 billion and HDI ranking is 145. Emissions of CO_2 per capita are 0.3 metric tons.

Flooding and Drought

Kenya is besieged by flooding and drought, which are both projected to increase considerably in the twenty-first century. The most common natural disasters since 1968 have been floods (48), epidemics (32) and droughts (14). Most of the deaths (4856) have been due to epidemics, which are a frequent consequence of preceding floods or droughts. Droughts account for most of the affected people (48.8 million).[5] In 2016, heavy rainfall battered several parts of the country, including the regions of Wajir, Marsabit and Turkana, which resulted in thousands of citizens being displaced. In Nairobi, the capital, 85 mm of rain fell in just three hours on April 28, 2016, causing widespread flooding. Moreover, three months of persistent and heavy rains from October to December, 2016, resulted in 112 fatalities and left 100,000 citizens displaced. The worst hit areas included Mt. Elgon, Kirinyaga, Narok, Busia, Kisumu, Tana River, Trans Nzoia, Busia and Bungoma counties. Severe infrastructure damage was evident in many parts of the country. The rains were exacerbated by a particularly strong El-Niño effect from the Pacific Ocean. In the same year, 23,000 people in Kwale and Turkana were devastated by persistent flooding and 6634 people in Taita Taveta.

A report on climate projections for Kenya reported the following scenarios: "Flood magnitudes in Kenya could increase with climate change, a tendency for (sometimes very large) increases in flood risk, a recent study provides new knowledge relative to the IPCC AR4, for coastal impacts in Kenya. A 10% intensification of the current 1-in-100-year storm surge combined with a 1m Sea Level Rise (SLR) could affect around 42% of coastal total land, 22% of coastal agricultural land, 32% of coastal GDP, and 39% of coastal urban areas."[6] The 536 km coastline of Kenya is extremely vulnerable due to sea level rise. As a developing country with a high poverty ratio and a per capita income of USD 1340 in 2016, many communities will struggle to adapt to coastal flooding and the consequent migration pressures.

Drought is a serious threat to the Kenyan population. As temperatures steadily rise over the coming decades and dry conditions persist, drought conditions and events will become more pronounced. In 1975, "widespread drought affected 16,000 people, in 1977 over 20,000 were affected and in 1980, 40,000 people experienced the effects of prolonged drought. Over 200,000 people were affected by drought in 1983/84 and in 1991/92, in arid and semi-arid districts of North-Eastern Kenya, the Rift Valley, Eastern and Coastal Provinces, 1.5 million people were affected by drought. From 1969 to 2004, 165 died due to drought, while 16,312,600 people were affected. Studies indicate that counties including Turkana, Kitui, Garrissa and Tana River experience erratic rainfall and prolonged drought in the recent past 9 of the 14 droughts reported since 1964 have occurred since the beginning of the 1990s."[7]

The drought in 2014 caused hardship to an estimated 1.6 million Kenyans and conditions persist in 2017. This of course has a major impact upon the agriculture sector, employment and food security. In Kenya, the drought cycles have become shorter, more frequent and intense due to global climate change and environmental degradation, and as the cycle becomes shorter, more and more people get affected and the impact gets even more severe.[8] As with China, the Kenyan population is facing a crisis of desertification that is exacerbated by drought. It is estimated that approximately 23% of land in the country is vulnerable to soil erosion and desertification.

Extreme Weather

Extreme weather events are a climate phenomenon that is occurring with more regular frequency. Environmental data over the previous five decades indicates a clear pattern of volatile weather events. The concurrent human and environmental costs of these events are a serious burden to the developing country. "Since 1968, Kenya experienced a total of 101 natural disasters (droughts, floods and related epidemics) with a total of

58.66 million people being affected and 6509 deaths. The majority of these events occurred over the past two and a half decades. This century alone has seen 72 events, and in the 1990s, a total of 19 events occurred, whereas the 1980s saw only 3 disasters."[9] In March 2017, Kenya faced another severe drought with 4 million people affected and food shortages causing hardship to 1.3 million citizens. Indeed, one of the tragic symptoms of climate change adversities and extreme weather is the disruption to agriculture and general food security. The effects of La Niña have also been devastating for Kenya with pronounced drought, low rainfall levels and flooding occurrences. The drought conditions and low rainfall patterns are particularly dangerous for a country that has approximately 65% of the land compromised by desertification or semi-desert conditions.

Data analyses on environmental trends in Kenya suggest disruptive and dangerous weather patterns between 2020 and 2100. "Average annual temperature will rise by between 1 °C and 5 °C, typically 1 °C by 2020s and 4 °C by 2100. Climate is likely to become wetter in both rainy seasons, but particularly in the Short Rain (October to December). Rainfall events during the wet seasons will become more extreme by 2100. Consequently flood events are likely to increase in frequency and severity. Droughts are likely to occur with similar frequency as at present, but to increase in severity. This is linked to the increase in temperature."[10,11] The government's 2010 National Climate Change Response Strategy noted with concern that drought cycles are increasing and with shorter intervals. Two severe droughts occurred in 1964 and 1984 then in 1996, 2004, 2006, 2007, 2008 and 2009. Both 2016 and 2017 have witnessed further drought crises and the resultant human catastrophe. As temperatures rise and dry conditions worsen, there is an alarming potential for annual droughts in Kenya.

Health, Human Development, Food Security

The population is projected to reach approximately 70 million by 2030. As population rises the concomitant impacts of climate change will affect even larger numbers of citizens and be significantly more challenging for vulnerable groups such as the impoverished, elderly, disabled and health challenged. One mitigating factor is that the population is relatively young with 13.7 million people under the age of 35 and citizens under 25 comprising 63% of the population. Yet overall, drastic climate and environmental crises remain for the foreseeable future, which will burden the population in terms of health and economics.

A serious health challenge in the country is malaria, which is exacerbated by climate conditions and thus expected to increase in intensity and frequency. Malaria has "a large negative impact on farm labor. Women and children are particularly vulnerable. Consensus is growing in Kenya that the malaria epidemic is connected to changing climate conditions. Highland areas, especially in East Africa, will likely experience increased malaria epidemics as temperatures increase and areas above 2000 m, with temperatures currently too low to support malaria transmission are affected.[12] Further analysis has affirmed that in Kenya in 1997/1998 and 2006/2007 for Wajir County in the north-east, extreme climate events were associated with a large malaria epidemic resulting in high admissions to Wajir Hospital and a weekly malaria incidence of 40–55 cases per 1000 population per week in all persons and children."[13] Climate change can also trigger negative impacts upon infectious disease levels. The spread of meningococcal (epidemic) meningitis is often linked to climate change, especially drought, and areas of sub-Saharan and West Africa are sensitive to the spread of meningitis and will be particularly at-risk if droughts become more frequent and severe.[14]

Changing environmental conditions will have a profound effect upon the agriculture sector and consequently health and food security. Agriculture represents 68% of employment. Approximately 85% of the land area is classified as arid or semi-arid and thus highly dependent on the rainy seasons. Changes in the duration and intensity of the dry season have a severe effect upon agriculture and cultivation. Thus the widely anticipated acceleration of drought in the country will immeasurably impact health and food security. Kenya is currently a country of moderately high levels of undernourishment and several global-scale studies project that Kenya could face increasingly serious food security issues over the next 40 years.[15]

New research suggests that several agriculture sectors will be negatively impacted by environmental changes. For example, bean yields in East Africa are estimated to experience a reduction by the 2030s under an intermediate emissions scenario (A1B)[16] and by the 2050s under low (B1) and high (A1FI) emissions scenarios.[17] Banana and plantain production could decline in lowland areas of East Africa, whereas in highland areas of East Africa it could increase with temperature rise.[18] It is also anticipated that a change in weather patterns could expand the distribution of ticks causing animal disease, in particular theileriosis (East Coast Fever) disease, which causes anemia and skin damage that expose cattle to secondary infections.[19] Moreover, ticks and tick-borne diseases will exacerbate the growing food insecurity among the pastoral community in Kenya.[20] The pastoral community is already struggling with the dual problems of poverty and inadequate nutrition levels. Pastoralists, who are nomadic tribal groups who live off the land, raising animals and cultivating crops, represent 60% of the population and utilize 70% of the country's accessible land.

Climate Change Refugees

Climate change crises will generate significant migration pressure and potential external movement of people to neighboring countries in search of safety and security. The main push factors will be drought and flooding from sea level rise and tropical storm systems such as La Niña. In Kenya, a total of 58.6 million people have been affected by natural disasters since 1964, of which the great majority, 48.8 million, was due to droughts and only about 3 million due to floods.[21] The most seriously impacted drought counties are those mainly populated by pastoralists, namely Baringo, Laikipia, Turkana, Samburu, Narok, Kajiado, Marsabit, Isiolo, Mandera, Garissa, Wajir, Tana River, Kilifi, Kwale and Taita Taveta.[22] Flooding displaced 180,300 people in 2013, and 97,000 in 2012. It is well established by numerous reports and projections that the majority of climate change refugees will be generated on the African continent. East Africa, where Kenya is situated, will experience a high proportion of these unfortunate refugee pressures. A UN conference on migration noted that, "twenty five million people migrated for environmental reasons, a number that is expected to rise to two hundred million by 2050, mostly affecting women and children. Only the advancing desert in Sub-Saharan Africa already lead millions of people to migrate that region to north of the continent and to south of Europe, by 2020."[23]

City Profile: Mombasa

Mombasa, with a population of 1.25 million (2017), is situated on the east coast of Kenya bordering the Indian Ocean and serves as major seaport. It is located on Mombasa Island, which is linked to the mainland by the Nyali Bridge, the Likoni Ferry and the Makupa Causeway. The land area of the city is 229.9 km^2 with 200 miles of coastline. Serious climate change threats to the city include sea level rise, flooding, saltwater intrusion, storms, health crises and infrastructure damage. A 2014 UN report warned that the city and coastal area surrounding Mombasa will face severe flood threats by 2080, and indeed the potential for near submergence of Mombasa itself. The report also projected that nearly 30% of the African coastal infrastructure could be severely damaged by 2080 due to rising sea levels, which are a known certainty.

Mombasa faces a number of problematic socio-economic issues that will complicate climate change impact, particularly flooding. Moreover, a high rate of poverty makes thousands of citizens highly vulnerable to climate change crises. As noted in a World Bank report, "In Kenya's two major cities—Nairobi and Mombasa—water demand exceeds supply by more than 150,000 and 100,000 cubic meters per day, respectively. Only about 18 percent of the total urban population has access to a sewer, 70 percent rely on septic tanks and pit latrines, and the rest have access to no sanitation services at all. In addition, existing wastewater treatment systems operate at very low efficiencies (about 16 percent of design capacity for 15 plants assessed in 2010), leading to discharge of untreated effluents. No urban area in the country has a properly engineered sanitary landfill, and most solid waste is dumped in open dump sites or other undesignated areas, or burned."[24] Recent studies estimate that a sea level rise of only 0.3 m would submerge an estimated 17% of Mombasa, with extensive areas left uninhabitable due to flooding, or agriculturally unsustainable due to saltwater flooding.²⁵ The devastating floods of October 2006 affected 60,000 people and caused extensive health problems and infrastructure damage.

Innovation and Adaption Strategies

1. In 2016 Kenya ratified the Paris Agreement.
2. The 2016 Climate Change Act provides a regulatory framework for an enhanced response to climate change and a mechanism and measures to achieve low carbon climate development. The government also established the National Climate Change Council, chaired by the President of Kenya.
3. The government has set a target of universal access to electricity (currently 55%) by 2020.
4. The Mitigation Strategy has set a 30% emissions reduction by 2030.
5. The 2013 National Drought Management Authority Bill was drafted to establish NDMA.
6. The government projects a 20% growth in the solar photovoltaic sector leading to a 10 MW capacity generating 22 GWh annually by 2020. Solar-power-use goals include the installation of at least 350,000 solar water heating units by 2017 and a goal of 100 MW from solar power by 2017, 200 MW by 2022 and 500 MW by 2030.
7. There is a government target of a wind power capacity increase of 500 MW by 2017, 1000 MW by 2022 and 3000 MW by 2030.
8. The Biogas for Better Life program sets a target of 5000 bio-digesters by 2017, 6500 by 2022 and 10,000 by 2030.
9. In 2010, the Ministry for Environment and Mineral Resources launched the National Climate Change Response Strategy (NCCRS), complemented by the 2013–2017 Climate Change Action Plan.
10. In 2010 a new constitution was adopted. From an environmental and health perspective, Article 43 is significant as it guarantees the right to the "highest attainable standard of health," the right to "accessible and adequate housing," the right to "adequate food of acceptable quality" and the right to "clean and safe water." The "right to a clean and healthy environment" is codified in Article 42.

11. According to Article 69(1), the state will (a) ensure sustainable resource use and exploitation, and "the equitable sharing of the accruing benefits"; and (b) "establish systems of environmental impact assessment, environmental audit and monitoring of the environment."
12. In 2005 Kenya ratified the Kyoto protocol.
13. In 1994 the UNFCCC was ratified.

South Africa

South Africa is a leading economy and developing democracy in Africa. The country has a population of 57 million (2017), a land area of 1.22 million km^2 (470,693 square miles) and an extensive coastline of 2798 km. Bordering nations include Botswana, Lesotho, Mozambique, Namibia, Swaziland and Zimbabwe. Life expectancy is 53 years (men), 54 years (women) and the country has an HDI ranking of 116. The GDP (purchasing power parity) was USD 723.5 billion in 2015. In 2016, its contribution to CO_2 emissions was 1.16%. The country faces significant impact from climate change including sea level rise, saltwater intrusion, flooding, drought and elevated health challenges in a nation already experiencing high rates of poverty (20%) and unemployment (30%). The poverty stricken, unemployed, health challenged and those living in slums and informal settlements are extremely vulnerability to climate change adversities. Temperatures in South Africa are steadily increasing with a projected rise of 1–2 °C in coastal regions and 3–4 °C in the interior by 2050. By 2100, the projections are 3–4 °C in coastal regions and 6–7 °C in the interior, which will exacerbate water insecurity and drought conditions. The coast faces imminent danger from sea level rise and the attendant crises that this phenomenon will trigger.

Flooding and Drought

As a coastal nation with a significant coastline, the country will face continued threats from tropical storms, sea level rise and flooding. Increasing rainfall patterns that are projected for South Africa also increase the risk of flooding, mudslides, infrastructure damage and human trauma. In 2016, the country was devastated by the effects of El Niño, which caused flooding and seriously damaged crops. An estimated 18 million people in the southern Africa region were subsequently reliant upon emergency food aid from the World Food Program. According to climate and weather projections based upon 21 model simulations comparing the period 1980–1999 with the period 2080–2099, the following results are anticipated: a 3.4 °C increase in annual temperature (up to 3.7 °C in spring), a 23% decrease in winter rainfall and a 13% decrease in spring rainfall.[26]

Rising temperatures, drought and water insecurity are inter-related problems that increasingly threaten the population and will do so for the foreseeable future. In April 2017, water levels in the Cape region became perilous as dam levels plunged to a 22% capacity with only 12% usable water remaining. Several regions including Stillbay have been impacted severely. In May 2017, the Cape Town municipal government stated that the dam levels were at 22.8% (storage levels) and 10% of the dam water was not usable. The consumption level of 680 million L is 80 million L above the target level. Further dry conditions, drought and water over-use will intensify water insecurity.

A report examining sea level rise (SLR) in the southern Africa region noted a myriad of concerns that also threaten South Africa. "The south-eastern coastline of southern Africa, comprising South Africa, Mozambique, Tanzania and Kenya, is regularly affected by cyclonic and other significant weather events that have the ability to unleash large wave events along the coast. Much of this coastline comprises sandy beaches backed by flat low coastal plains that are already vulnerable to flooding and erosion in extreme wave events. Progressive SLR will worsen the situation but it is the episodic wave events, occurring with little advanced warning that results in significant flooding and erosion."[27] Approximately 40% of the population live in coastal regions thus the human security threats from sea level rise and flooding will be immense. The major coastal cities of Cape Town, Durban and Johannesburg, for example, will be seriously challenged.

A DIVA analysis indicated that substantial flood and sea level rise risks are projected for Africa with attendant damage and human suffering on a large scale. Moreover, in South Africa, the high rate of poverty, a plethora of low sea level coastal cities and towns and prohibitive mitigation and adaption costs will render the challenge exceedingly difficult to manage. "Without adaptation, the physical, human and financial impacts will be significant. On an African scale, (a 43-cm rise) approximately 16 million people will be flooded per year in 2100, 10 million people will be forced to migrate from 2000 to 2100, and there will be a total damage cost of US$38 billion per year in 2100."[28] Significant damages will be incurred in the following countries. "For economic damages they are Algeria, Egypt, Morocco, South Africa, Tunisia, Libya and Cameroon who are all estimated to have more than US$1 billion of additional damage per year in 2100. In absolute terms, the highest adaptation costs occur in Mozambique, Guinea, Nigeria, Guinea-Bissau and South Africa."[29]

The other major climate challenge facing the country is drought, which is expected to increase in intensity in the twenty-first century. In 2015, South Africa was struck by the worst drought since 1982. Five provinces were declared disaster areas by the prospective governments: The UN reported that 11 million children were at risk of serious undernourishment and water scarcity. The national government provided subsistence farmers with ZAR 300 million (USD 21 million; GBP 14 million) in assistance. Moreover, agricultural employment plunged by 37,000 in the fourth quarter of 2015, which elevated the unemployment rate to 24.5%. In sum, one severe drought episode in 2015 caused widespread human, financial and employment crises. The projections are for increased drought episodes in the country and water insecurity over the coming decades.

Analysis on drought projections for South Africa are serious and cause for intensified mitigation efforts. "During the austral summer months, dry conditions are projected for the southwestern region of South Africa. Additionally, austral spring months are projected to be drier, implying a delay in the seasonal rains of subsequent summer months. Under all future climate scenarios considered by the LTAS, higher frequencies of drought events are projected. In southern Africa, there is medium confidence that droughts will intensify in the twenty-first century in some seasons, due to reduced precipitation and/or increased evapotranspiration."[30] Climate scientists have calculated that both minimum and maximum temperatures will rise by 2–3 °C over the interior by 2040–2060. "Significant increases of more than 4 °C are projected for the far future period of 2075–2095."[31] Several regions in the country are experiencing the repercussions of drought, including water scarcity, which impacts critical irrigation for agriculture and national food security. In 2015, the Western Cape Local Government Department issued water advisories and the Agriculture Department assigned ZAR 47 million to drought relief mitigation efforts.

Extreme Weather

Extreme weather events include storms, extreme heat, drought episodes and erratic rainfall patterns. All of these weather phenomena are anticipated in South Africa in the long term. Moreover, the effects of El Niño have been adverse upon the country, including elevated temperatures and below average rainfall levels. Over the 2005–2017 period a general warming trend has been recorded for most of Africa, including South Africa. Before 2015 the country was experiencing drought conditions which were exacerbated by the 2015 El-Niño effect. Extreme weather has a myriad of symptomatic effects on the population including flooding, heat extremes, heat related health crises, infrastructure damage, damage to soil, crops and future cultivation, increased disease incidents such as malaria, cholera and diarrhea, which is particularly dangerous for impoverished children.

An analysis of projected climate change scenarios in South Africa for three periods (2015–2035, 2040–2060 and 2070–2090) up to 2050 observed the following calculations. "Four fundamental climate scenarios at national scale, with different degrees of change and likelihood capture the impacts of global mitigation and the passing of time. (1) warmer (<3 °C above 1961–2000) and wetter with greater frequency of extreme rainfall events. (2) warmer (<3 °C above 1961–2000) and drier, with an increase in the frequency of drought events and somewhat greater frequency of extreme rainfall events. (3) hotter (>3 °C above 1961–2000) and wetter with substantially greater frequency of extreme rainfall events. (4) hotter (>3 °C above 1961–2000) and drier, with a substantial increase in the frequency of drought events and greater frequency of extreme rainfall events."[32] In 2017, examples of extreme rainfall events included February 22–23, 104 mm in 24 hours in Thabazimbi; 76 mm in 24 hours in Lydenburg; 88 mm in 24 hours in Marken; and for February 21–22 there was 111 mm in 24 hours at Richards Bay and 101 mm in 24 hours at Mutunzini.

The government of South Africa is moving to address the increasing challenges of extreme weather events that according to numerous scientific calculations will be more pronounced in the twenty-first century. As noted in one government report, "Climate change will require more effective disaster management to deal with the increased number of extreme weather events. The increase in extreme events will strain public resources due to the need to declare and support disaster areas in an immediate crisis as well as during long-term recovery. In response to these challenges, South Africa will: continue to develop and improve its early warning systems for weather and climate (especially severe weather events)."[33]

Health, Human Development and Food Security

In 2016, the level of poverty was 15.9% and the unemployment rate was 27.3% in the third quarter, which posed substantial health, social and economic burdens on the population. Adverse climate change impacts traditionally fall hardest on marginalized, health compromised and impoverished groups in society. Additionally, these are the groups who are least able to personally mitigate climate change effects such as seeking medical attention, repairing damaged dwellings or migrating from weather damaged regions. In 2016, the age demographic profile of the country was: 0–14 years: 28.34% (male 7,718,511/female 7,667,830); 15–24 years: 18.07% (male 4,865,807/female 4,943,707); 25–54 years: 41.44% (male 11,372,944/female 11,130,874); 55–64 years: 6.59% (male 1,662,874/female 1,915,908); 65 years and over: 5.57% (male 1,269,551/female 1,752,698). One encouraging relief factor is that the population is relatively young (78% under the age of 55) thus indicating in some instances a better coping capacity for environmental challenges.

Africa: South Africa

> A leading international assessment of the effects of climate change on the global economy, the Stern Review, estimates that damages from unmitigated climate change could range between 5 and 20% of global GDP annually by 2100. In the absence of effective adaptation responses, such levels of damages would certainly threaten and even reverse many development gains made in South Africa.[34]

There are multiple and serious health complications from climate change that will affect the population. These include respiratory problems, heat stroke, disease, unsanitary conditions caused by damage to water systems and sewage systems, cardiovascular complications and vector-borne infectious diseases such as malaria. A government white paper on climate change noted that, "a significant proportion of South Africans, and in particular the poor, already have serious and complex health challenges compounded by poor living conditions. These include amongst the world's highest rates of tuberculosis and HIV infection. In particular parts of the country, the coverage of vector-borne diseases like malaria, rift valley fever and schistosomiasis may spread due to climate change, requiring a concomitant expansion of public health initiatives to combat these diseases."[35]

In terms of food security for a nation that already has a poverty level of 15.9%, climate change will have considerable adverse effects. One of the projections for South Africa and the African continent in general is that food prices will escalate to unsustainable levels, thus posing additional hardships on impoverished populations battling chromic malnourishment and insufficient caloric intake. The agriculture sector will be impacted by drought conditions and water insecurity. This in turn will affect crop cultivation, employment and soil quality. "South Africa is now in its second year of below optimal crop production conditions with the current season being the worst on record. The country's average rainfall for 2015 was the lowest since 1904, according to the South African

Weather Service. Temperatures and the amount of rainfall (intensity and distribution) are critical for agricultural production and the security of domestic food supplies."[36]

An analysis of climate induced effects on agriculture in South Africa forecasted serious problems due to enhanced levels of drought. "South African agriculture is divided into three main subsectors, namely livestock (which accounts for 46% of the gross producer value), field crops (28%) and horticulture (26%). The impact of the drought differs across sectors, but it is particularly severe in the two largest subsectors due to their dependence on dry land grain production. Livestock under extensive production systems has been severely affected as it is largely dependent on grazing. Some producers have lost their stock, particularly in the emerging sector, while others had to cull and reduce their herds."[37] Moreover, the government has acknowledged the critical issue of food security relative to climate change and cited three emerging crisis scenarios. "South Africa might become a net importer of food and this will negatively affect the trade balance." First, sectors such as maize and sugar, which would normally contribute to the sector's positive trade balance, will shift to a negative net trade position. The reduced production volumes and planted areas in the case of grains and oilseed crops will result in serious financial losses for input suppliers due to lower demand for seed, fertilizers, pesticides and herbicides. Second, at a food manufacturing level, the shortage of grains and higher prices will result in higher production costs. Moreover, maize has to be imported due to reduced availability domestically. Third, it is estimated that over 2.9 million households continue to depend on agricultural production and their finances and food security will be severely affected.[38]

Climate Change Refugees

With an extensive coastline South Africa will be impacted in devastating ways by sea level rise, flooding, saltwater intrusion, infrastructure damage and health crises. These climate phenomena will create massive migration pressures away from coastal regions where 40% of the population reside. A major report on sea level rise, which substantially affects South Africa, stated that, "sea-level rise projected for 2100 poses significant threats to coastal zones in the world. Particularly, when the intensification of tropical cyclones is superposed on sea-level rise, the population at risk from inundation is likely to amount to several hundred million. We should therefore consider that climate change is an issue for security on how to guarantee the safety of such a large number of people, including those in Asia and Africa."[39]

South Africa will definitely experience significant internal climate refugee flows in the twenty-first century and may also have refugee movements to neighboring countries, although this eventuality will be dependent upon a number of factors related to food and economic security. The Intergovernmental Panel on Climate Change report (IPCC 2014) projected a pronounced increase in climate change refugee flows and pressures in southern Africa. A report by the Migration Policy Institute noted that in 2015, Angola, Botswana, Malawi, Namibia, South Africa and Zimbabwe experienced extreme droughts with large impacts on both staple foods and cash crops, which will leave many with decreased means of sustenance or income, and may motivate migration to urban areas or across borders.[40] These strong climate related migration pressures are expected to increase notably. Drought, food insecurity and coastal flooding are three powerful factors that will generate considerable climate refugee flows in future decades.

Innovation and Adaption Strategies

1. 2016 South Africa signs the Paris Agreement;
2. the Constitution of the Republic of South Africa National Environmental Management Act (NEMA);
3. National Environment Management Bio-diversity Act (NEMBA);
4. 2008 National Bio-diversity Framework 2008 (NBF);
5. National Environment Management: Air Quality Act (NEMAQA);
6. National Environmental Management: Protected Areas Act (NEMPAA);
7. National Environmental Management: Integrated Coastal Management Act (NEMCMA), National Environmental Management: Waste Act (NEMWA);

8. National Water Act (NWA);
9. National Forest Act (NFA);
10. 2002 Kyoto Protocol ratification (July 31, 2002);
11. 2000 white paper on Integrated Pollution Waste Management for South Africa;
12. 1998 white paper on Environmental Management Policy for South Africa;
13. 1998 white paper on the Energy Policy of the Republic of South Africa;
14. 1997 white paper on the Promotion of Renewable Energy and Clean Energy;
15. UNFCCC: signature (June 15, 1993) and ratification (August 29, 1997);
16. 1990 Vienna Convention: accession (January 15, 1990);
17. 1990 Montreal Protocol: accession (January 15, 1990).

Botswana

Botswana is a poor, landlocked country with a population of 2.1 million people and GDP of USD 10.9 billion. The land area is 581,730 km^2 (224,607 square miles). It borders Namibia, South Africa, Zambia and Zimbabwe. Life expectancy is 54 years (men), 51 years (women) and its HDI ranking is 106. Approximately 19% of the population live in poverty, 17.8% are unemployed and the HIV/AIDS rate is a staggering 22%. These daunting human development factors limit personal and national responses to effective climate change mitigation and adaption strategies. In general the climate is semi-arid with warm winters and hot summers. The main environmental challenges are desertification, overgrazing, drought and water scarcity. Moreover, Botswana is projected to be one of the first countries to experience extreme water scarcity that will significantly undermine national socio-economic development. The 2015 El-Niño effect directly influenced water security and agriculture with a 70% reduction in cereal harvests, drought episodes and emergency funding from the Ministry of Agriculture with subsidies on certain livestock feeds increased by 50%.

Flooding and Drought

Flooding and water damage are recurring problems in the country though have not reached crisis proportions. In March 2017, flooding caused serious damage and disruption in Gaborone and nearby areas. In February 2017, Botswana was struck by a tropical depression, ex-Dineo, which caused extensive flooding across Botswana. Serious infrastructure damage to schools, roads, bridges and homes was reported. Moreover, the Gaborone Dam, which is the primary water course for the capitol of Gaborone, reached over-capacity. During the 2016–2017 rainy season, the country recorded rainfall levels of between 110 and 220 mm. As noted by a UNFCCC report, "Botswana is vulnerable to the impacts of climate change, the assessment from the Second National Communication indicate that rainfall has been highly variable, spatially, inter and intra annual and that droughts in terms of rainfall deficits are most common in northern Botswana. Extreme droughts based on low rainfall and soil conditions are most common in south-western Botswana and high rainfall events with risks of floods are most likely in northeastern Botswana where several large dams are located in this area. Droughts are projected to increase in frequency and severity. Botswana is already witnessing impacts of climate change with constrained agricultural production, increasing food insecurity and increasing water stress, which will worsen with time, as projected."[41]

Data provided by the National Disaster Management Office indicated that due to the February 2017 flooding, "650 households had been severely affected as a result of ex-Dineo, over 500 houses have been destroyed and infrastructure, telecommunication lines and livelihoods have been disrupted in the affected districts. The situation resulted in moderate population displacement, which required the hosting of some 300 households in community halls, schools and churches. Water sources have also been negatively affected, which poses a further threat to the displaced. Destroyed latrines, stagnant water, and contaminated boreholes have heightened the healt research is also under way h risks as well as the risk of waterborne and communicable diseases."[42] Climate change impacts will expose the country to increasing levels of flooding and subsequent health and economic hardship.

A more serious and immediate threat facing the country is drought. Although this climate condition is an on-going problem, reliable projections indicate that the crisis will mount in intensity over the twenty-first century. An IPCC report stated that in a future scenario of 2040–2059, "a drying trend in western portions of the country and surrounding region (extending into desert areas of Namibia and Botswana) is projected to continue through the century. The southwestern regions of the country are thought to be at high risk of severe drought during this century and beyond.[43] As noted in an agricultural assessment of climate impacts, the trend of dry weather and drought conditions has already intensified in clear and measurable terms. "Rainfall was poorly distributed and below average at the beginning of the cropping seasons for four years between 2013 and 2017. For 1990, 2003, and 2012, the standardized precipitation index (SPI) was −1.77 (severe drought), −1.37 (moderate drought), and −2.32 (extreme drought), respectively. To minimize impacts on crop production, farmers simultaneously planted different crops based on the perception that climatic impacts on different crops vary and favored crops perceived as drought resistant. Livestock farmers supplemented with livestock feeds, reduced herd size, and moved livestock to areas with better forage. Off-farm incomes from selling products harvested from the wild are important to farmers as a coping strategy, particularly when rain fails."[44]

Combined future impact of climate change upon Botswana and the region are daunting for human health, economic and infrastructure management. As noted in IPCC studies addressing future climate scenarios for South Africa, Lesotho, Swaziland, Namibia, Angola, Botswana, Zimbabwe, Zambia, Malawi, Mozambique and Madagascar there will be significant effects. These include:

"Increase of 0.6–1 °C over most of the region; Projected Climate Change Temperature: Mid-term (2046–2065): increase of 2–3 °C, evenly distributed across the region; Long-term (2081–2100): increase of 2–5 °C, slightly warmer inland (Namibia, Angola, Botswana); The warmest daily maximum temperature is projected to increase 4–6 °C, evenly distributed across the region; The number of tropical nights (that is the number of 24-hour days above 20 °C) is projected to increase from between 90 and 100 days; South Africa is projected to have slightly fewer but the rest of the region is projected to have 100 days of warm tropical nights; Precipitation: Mid-term (2046–2065): the largest changes in precipitation appear to occur in June–August and range from a 10 to 30% decrease, with most of this drying occurring in Namibia, Angola, Botswana, and Zambia; Long-term (2081–2100): 20 to 50% decrease in June–August, mostly on the western coast of the region (Namibia, Angola, Botswana, Zambia, South Africa; Overall, annual precipitation is projected to decline across the region by the end of the century, especially in Mozambique; Decrease in cloudiness and humidity across the land masses of the region; Decrease in annual mean soil moisture of −1.2 mm to −2 mm across the region; Decrease in annual runoff of up to −30 in the western portion of the region (Namibia, Angola, Botswana, Zambia, South Africa)."[45]

Extreme Weather

The country faces extreme weather events in the form of drought, erratic rainfall patterns and windstorms. This in turn has an effect upon health, agriculture production, employment, water scarcity and desertification. The Gaborone Dam has witnessed a capacity drop from 141 million m^3 to approximately 4 m above ground level (1.7% capacity) in 2016 and a rebound to 71% capacity in 2017, which highlights the erratic nature of its water levels. Moreover, about 60% of water usage is derived from surface water, which places further strains upon water sustainability. As noted in an African Development Bank report on Botswana, a looming water crisis faces the country. "With rainfall ranging from a high of 550 and low of 200 mm per year, and an estimated annual average evaporation rate of 1400 mm, Botswana is a water stressed country. The bulk of the country i.e. west and southwest, is occupied by the Kalahari and only the northern and eastern rim of the country enjoy any form of vegetation. The rain falls in the summer months of November to April."[46]

The government is taking steps to address the drought situation in the country and the corresponding lack of water security that threatens the population. As climatic extreme events are cross-cutting and affecting all economic sectors, the government has adopted a strategy that encompasses all economic sectors with an emphasis on the water, health and agriculture (crop and livestock) sectors.[47] The country is also faced with a record of natural disasters that exacerbate socio-economic challenges and climate change mitigation. "Botswana is among the countries in the world with the highest number of people affected by natural disasters (13,529 per 100,000 inhabitants) [in] the last three decades. All 11 natural disasters between 1974 and 2003 are hydro-meteorological disasters, of which 7 are related to drought, 3 to flooding and 1 to windstorm disasters. More than 1 person out of 10 is affected in Botswana. Of the total number of victims (13,529), over 93% are affected by droughts, 2.5% by floods, and 4% by windstorms."[48]

Health, Human Development, Food Security

As a small, impoverished and landlocked country with a rural population of 963,000 (42.5%) in 2015, the negative impacts of climate change will pose considerable hardships upon the population. Approximately 70% of rural inhabitants derive their income from the agriculture sector, thus changes in temperature and drought conditions have a profound impact upon livelihoods and affect multiple social, health and economic indicators. "Botswana rainfall is low, highly erratic and unevenly distributed. Surface and ground water resources are scarce. Water resources are very limited and are expected to constrain future economic growth if not used efficiently. The country currently faces acute water shortages in south eastern Botswana and its capital Gaborone, which have led to severe water restrictions and supply interruptions. Climate change is expected to exacerbate the situation, leading to more droughts (and floods), increased stress on water resources and reduced primary land productivity."[49]

A UNFCCC summary of the more significant climate change impacts facing the country in the twenty-first century underscores the severity of the challenges that will be experienced. "In Botswana, under a high emissions scenario, mean annual temperature is projected to rise by about 6.2 °C on average from 1990 to 2100. If global emissions decrease rapidly, the temperature rise is limited to about 1.7 °C. In Botswana, under a high emissions scenario, the number of days of warm spells is projected to increase from fewer than 10 days in 1990 to about 240 days on average in 2100. If global emissions decrease rapidly, the days of warm spells are limited to about 50 on average.[50]

Climate change induced drought, heat waves and flooding will also create health crises for the population. Two symptomatic risks are heat stroke and elevated levels of malaria, which can occur in hotter climates and which flood conditions may exacerbate. The risk in Botswana of malaria "is projected to increase under a high emissions scenario. In Botswana under a high emissions scenario, heat-related deaths in the elderly (65+ years) are projected to increase to about 136 deaths per 100,000 by 2080 compared to the estimated baseline of approximately 3 deaths per 100,000 annually between 1961 and 1990. A rapid reduction in global emissions could limit heat-related deaths in the elderly to about 20 deaths per 100,000 in 2080."[51] Moreover a report by the INDC attests to multiple climate change events that seriously threaten the population and economic stability. "Droughts and rainfall variability are predicted to increase with climate change, slowed agricultural production, increasing food insecurity and increased water stress have already been witnessed, and are likely to continue. Extreme events associated with climate change are likely to lead to an increased incidence of vector-borne diseases such as malaria and Bilharzia."[52]

Climate Change Refugees

As a small landlocked country with substantial drought projections in the twenty-first century as well as water scarcity and food insecurity, it is expected that Botswana will generate internal climate refugee flows. In turn, this will likely generate external environmental refugee movements to neighboring countries. In 2015, the Botswana government allocated emergency funding to provide relief to livestock farmers in response to the worst drought in nearly four decades. The undermining of rural livelihoods in farming and herding and increased competition for resources may contribute to migration decisions.[53] As 70% of the population earn income from farming it is evident that drought episodes, which will increase in the twenty-first century, food insecurity and the looming water scarcity crisis will compound migration pressures for farmers and their families.

A study examining the agriculture sector observed that, "Botswana is a semi-arid, middle-income African country that imports 90 percent of its food. Despite its relative prosperity, Botswana also suffers from one of the highest measures of income inequality in the world, persistent poverty, and relatively high levels of food insecurity."[54] One likely scenario is that rural climate migration will occur in large numbers to the capital city of Gaborone, which already faces water insecurity challenges. Other smaller cities may also face migration inflows. Depending on the severity of drought conditions and food insecurity, a potential also exists for Botswana citizens to migrate as climate change refugees to neighboring countries, or indeed further destinations including Europe. Many national security analysts have noted that the burgeoning climate change refugee crisis, which will accelerate to between 200 million and 500 million refugees by 2050, will increase pressure upon Western industrialized states to erect and maintain militarized borders. Whatever the exact outcomes, it is a certainty that one of the tragedies of climate change adversity will be tens of millions of citizens, young and elderly alike, being forced to seek safety and sustenance far from home.

Innovation and Adaption Strategies

1. 2016 Botswana ratified the Paris Agreement;
2. 2015 commitment to reduce overall carbon emissions by 15% by 2030, INDC;
3. 2015–2030, the government of Botswana initiated six important sustainable goals over a 15-year time period as follows:

   - improving agricultural productivity and farmers' incomes;
   - building resilience and mitigation efforts;
   - value chain;
   - research for development and innovations;
   - improving agriculture advisory support services;
   - improved institutional coordination.

4. 2003 Kyoto Protocol ratified;
5. 1994 UNFCCC ratified;
6. 1986 Declaration of Forest Reserves;
7. 1971 Atmospheric Pollution Act.

Interview

Anne Nyatichi Omambia is an environmental scientist and climate change mitigation expert working as a compliance officer and climate change coordinator for Kenya's environment agency. She has 14 years of senior experience in environmental management and climate change research.

Can you describe your work related to climate change?

NEMA is a state corporation established by an act of parliament— Environmental Management and Coordination Act CAP 387.

We are in charge of all matters related to the environment as an environmental regulator.

Specifically with regard to climate change, I am the climate change coordinator in charge of all climate related functions of NEMA. These functions are:

1. Designated National Authority for the Clean Development Mechanism;
2. National Implementing Entity for the Adaptation Fund and the Green Climate Fund;
3. the Climate Change Act, 2016 assigns NEMA the role of enforcement of the Act and monitoring of GHGs. This includes ensuring that climate change considerations are incorporated in the Environmental Impact Assessment processes;
4. NEMA also implements The National Climate Change Action Plan 2013–2017 with the key role being environmental awareness, capacity building and public participation in climate change actions.

What concerns you the most about climate change in your country?

Kenya is a climate vulnerable country and thus my greatest concern is building and enhancing citizen resilience to climate hazards, especially vulnerable communities, and enhancing the ecosystem's resilience to sustain a growing economy and population along a low-carbon sustainable development pathway. The measures to achieve this are spelled out in our climate change policy documents. However, the greatest burden is the means of implementation.

What concerns you the most about climate change globally?

I am concerned that the world has the science and knowledge on climate change and what needs to be done, but countries are slow and sometimes hesitant to implement the quick win actions (politicizing climate change).

Can you describe a specific sustainability project in your area or country that is contributing to a "greener world" and lower carbon footprint.

Kenya has various sustainability projects across various sectors. At NEMA, we are the entity in charge of training government ministries and state corporations and agencies on sustainability reporting for purposes of annual government performance contracting. As a project, NEMA has built seven green buildings (Green Points) in seven counties to serve as an example to citizens on sustainable building. Further, NEMA has a sustainability policy and reports annually on its performance as part of performance contracting.

What steps could your government take to mitigate the growing threats from climate change?

Kenya has spelt out its key mitigation actions in its Intended Nationally Determined Contributions to the UNFCCC, the Kenya National Action Plan 2013–2017 and the National Climate Change Policy Framework. These mitigation actions cut across all sectors and if implemented can lead Kenya towards a low-carbon climate-resilient pathway.

What is your reaction to the UNIPCC's accepted calculation by Prof. Myers of a minimum of 200 million climate change refugees by 2050 and the world's readiness to cope with this humanitarian and environmental challenge.

The number of climate migrants will definitely steadily increase annually under the current business as usual scenario on a global scale. Even though I have not read the full IPCC report, the world needs to prepare for large numbers of climate change migrants and refugees and the associated resource conflicts that may arise.

Can you describe a particular area of climate change that you are concentrating upon with research, or a grassroots development project?

NEMA as a National Implementing Entity for the Adaptation Fund is implementing an adaptation program in various counties. The program document can be downloaded from the adaptation fund website (www.adaptation-fund.org) and our NEMA website (www.nema.go.ke).

On a personal level, I am involved in research in both adaptation and mitigation and have published with Springer journals. My author name is Anne Nyatichi Omambia.

Interview

Moctar Dembélé is a young researcher in water resources management. He has experience in irrigation systems and agricultural drought assessment. Currently, he is completing a Ph.D. in geosciences and environment to quantify the current and future water resources of the trans-boundary Volta river basin in West Africa.

Can you describe your work related to agriculture and drought and the work of the Africa Rice Center.

I started working for the Africa Rice Center as a junior research fellow after graduating as a water management engineer in 2014. I did my engineering training at the International Institute for Water and Environmental Engineering (2iE) in Ouagadougou, Burkina Faso. My work at the Africa Rice Center in Cotonou, Bénin, consisted of mapping drought in rice growing environments. I used Geographic Information Systems (GIS) and remote sensing to analyze the spatiotemporal distribution of agricultural drought in Burkina Faso. I used remotely sensed data to compute drought indices with agrometeorological data and applied time-series analysis on the period 2001–2013. After validating the methodology in Burkina Faso, I have applied it to nine other countries in Africa. The Africa Rice Center is one of 15 international agricultural research centers that are members of the Consultative Group for International Agricultural Research (CGIAR) consortium. It is committed to grow as a pan-African center of excellence for rice research, development and capacity-strengthening. Its mission is to contribute to poverty alleviation and food security in Africa, through research, development and partnership activities aimed at increasing the productivity and profitability of the rice sector in ways that ensure the sustainability of the farming environment.

What concerns you the most about climate change impacts on food security in Africa?

Africa is increasingly affected by the impact of climate change. It is known that people who are already vulnerable and food insecure are likely to be the first affected by climate change impacts. In Africa, climate change has been affecting all four dimensions of food security: food availability, food accessibility, food utilization and food systems stability. However, food availability is the most crucial as it conditions the other dimensions of food security. In fact, Africa is prone to recurrent extreme weather events such as droughts and floods that increase in frequency and intensity. The production of staple crops is affected by the erratic distribution of rainfall and the poor water control in floodplains that affect both crop quality and quantity, while high temperatures have an impact on crops yields. Natural disasters cause considerable losses, both economically and socially. Those events damage properties, take lives, deteriorate livelihoods and exacerbate poverty. Africa needs relevant development politics to face the challenge of climate change in order to alleviate poverty. Africa must be able to feed itself and at the same time use its natural resources in a more sustainable manner.

What concerns you the most about climate change impacts in Burkina Faso?

The geographical position of Burkina Faso makes it particularly vulnerable to climate change. As a landlocked Sahelian country in the heart of West Africa, Burkina Faso suffers from extreme and variable climates. Droughts and floods can occur in the same area with only a few months difference. In Burkina Faso, water shortages are becoming more and more recurrent, in part because water demand increases due to population growth and the expansion of water related activities, and because of climate change and contamination of water supply. Two extreme events demonstrated the fragility of Sahelian natural resources. These were the droughts of 1972–1973 and 1984–1985, which caused loss of human life and livestock in the semi-arid region. In Burkina Faso, flooding and drought affect agricultural lands each year and cause considerable economic losses. Severe droughts caused a significant decline in agricultural production, accentuating food insecurity of rural populations and leading to a slowdown in the growth of added value in agriculture. In fact, agriculture in Burkina Faso is dependent on rainfall that is characterized by its erratic distribution. The population is therefore at risk of starvation, and food insecurity threatens more. Agriculture is the mainstay of the economy as it contributes 35–40% to the GDP and employs 80% of the workforce of the country. Thus, Burkina Faso needs to strengthen its agriculture by developing key strategies for mitigation and adaption to climate change.

Can you describe a specific project to promote agriculture and mitigate drought in Burkina Faso that will have a positive impact?

Many initiatives have been implemented for climate change mitigation and adaption in Burkina Faso. The National Adaptation Programmes of Action (PANA) implemented between 2009 and 2013 was one of them. Its main objective was to implement urgent and immediate adaption measures to combat the effects of climate change, particularly for the marginalized and most vulnerable populations. The project focused on capacity building and human security improvement to mitigate and adapt to climate change. The activities and results included: training national experts on climate scenarios; conducting vulnerability assessments in agriculture, environment, livestock production, energy, health, infrastructure and natural disasters; evaluating economic impact and cost of adaption; and formulating an adaption program by 2025 and 2050. Another initiative is the CGIAR Research Program on Climate Change, Agriculture and Food Security (CCAFS) that started in 2011. It aims to improve interactions and communication among scientists, policymakers and civil society in order to entail changes in behavior, technology, institutions and food production systems. In Burkina Faso, CCAFS projects include the establishment of a climate-smart village (CSV) in Tougou in the Yatenga province located in the north of the country. The goal of the CSV is to define and implement portfolios of practical adaption technologies to improve food security and resilience to climate change impacts. Specific activities address gender related issues as well as the evaluation of climate smartness of technologies and practices. There might be gains in mitigation as well as in the chosen technologies that are designed to reduce GHG emissions. Another key activity is the climate-smart services that are being tested by farmers. Those services include tailored weather forecasts to plan planting, harvesting and other farm activities. Mobile phones are used to deliver advisories and weather forecasts, and they are also used to enable farmers to buy index-based insurance that gives them a measure of protection in the event of extreme weather. In addition to farm practices, a network of national stakeholders—including scientists and policymakers—regularly exchange knowledge on adaption to climate change through the national science-policy dialogue platform.

In Burkina Faso, who is harmed the most by the impact of climate change, for example farmers or rural people?

Female farmers are most affected by the impact of climate change in Burkina Faso. According to population analysis, they represent 52% of the labor force in the agricultural sector. However, they have limited access to resources and extension services such as land tenure, technology and micro-credits. Farmlands cultivated by women, either in groups or individually, are very often of poor quality and vulnerable to climate change. Women are heavily dependent on natural capital for their livelihoods so land degradation caused by climate change affects their livelihoods more drastically. Since they do not own the lands, women do not invest in them, or invest very little. Less physical strength and technical support mean that women rarely practice adaption techniques such as zaï (small holes or planting pits) or contour bunds to slow down water run-off, conserve water on plots to recharge groundwater, and prevent soil erosion. Fertilizers are reserved for family farmlands so that women's farmlands are poorly fertile, produce low yields and are more vulnerable to climate change. Moreover, beyond working in the field, women are also responsible for the well-being of the household. Thus, the increase in time and quantity of work is a blatant impact of climate change on women's human capital. Indeed, women spend more time fetching water, harvesting crops, processing grains, preserving and processing non-timber forest products, as well as collecting wood—which is increasingly scarce because of desertification and overexploitation. Women's participation in community life and engagement in income-generating activities is therefore limited by the time spent on their household duties. Moreover, girls are often forced to absent themselves from school to fetch water or relay the responsibilities of their mothers who do not have the time to do everything. This lack of time affects food quality, children's education, sanitation, health and the reproductive capacities of families.

What is your reaction to the UNIPCC's accepted calculation by Prof. Myers of a minimum of 200 million climate change refugees by 2050 and the world's readiness to cope with this humanitarian and environmental challenge?

In the long run, the number of people displaced by climate change will depend on our ability to deal with climate change impacts and develop adequate mitigation and adaptation strategies. The calculation by Prof. Norman Myers of a minimum of 200 million climate refugees by 2050 is stunning and daunting at the same time. It has become the accepted figure cited in IPCC publications and many climate change reviews. However, this figure should be analyzed further because even if the scientific argument for climate change is increasingly confident, it is difficult to know with certainty the impact of climate change on human population distribution. According to the UN, 20 million people were displaced by climate change in 2008. We can imagine that the estimates of 200 million climate refugees may be reached in 2050 if appropriate actions are not taken to firmly address the issue of climate change. However, the figure takes into account all of the people who have been displaced by any kind of natural disaster, and it is an overestimation that all of those single displacements are attributable to climate change alone. Moreover, in the case of natural disasters, people generally move away relatively short distances and then move back after a short period of time. Regarding the estimate of 200 million climate refugees by 2050, some researchers doubted the figure from its release. They argued that the method deployed by Prof. Myers to make his projections is controversial. Seemingly, Prof. Myers simply assumed that all the people affected by sea level rise in the world would have to migrate, and they were mostly in developed countries. If that is confirmed, then this basis is not strong enough. Prof. Myers was interviewed by the BBC and said: "It's really difficult to say how many there are and where are they…but in the long run I do believe very strongly that it will be better for us to find that we have been roughly right than precisely wrong." In his research, he also makes clear that not everyone he classes as an environmental refugee will flee their country because they could be forced to move somewhere else within national boundaries. Based on this assumption, the 200 million figure of climate refugees by 2050 is not actually a very large one if suitable mitigation strategies are not deployed in time, and if we consider the rapid growth in population, and the increase in natural disasters worldwide. A contentious issue is the definition of people displaced by climate change. There is a need to find a globally accepted definition applicable under international law with the commitment of the international community. The definition should account for the seriousness and the urgency of the situation, without denaturing the problem and reducing willingness to cope with the issue.

What agricultural crops in Burkina Faso are most at risk due to climate change?

Climate change might exacerbate the losses in crop yield, particularly for tropical cereal crops. Sorghum, millet, maize and rice are the major crops cultivated in Burkina Faso. According to a study of the CCAFS program, model-based estimates indicate that if no adaption actions are taken, on average, maize productivity could decrease by 5–10%, whereas rice productivity could decrease by 2–5% with every degree of warming during the twenty-first century in tropical areas. Moreover, most of the currently cropped maize area is projected to experience negative impacts, with production reductions in the range 12–40%. Previous studies in Africa conclude that maize (−5%), sorghum (−15%) and millet (−10%) yields are set to decline significantly in the future. In fact, projections highlighted optimal and marginal maximum temperatures for crop growing being exceeded, and a decrease in rainfall resulting in a deficiency of soil moisture required to meet crop water needs. In Burkina Faso, crop production is also constrained by land degradation due to soil erosion and infertility. Adaption of crop production to climate change should be a priority for the country. A major challenge for researchers is to develop technologies that lend greater resilience to agricultural production under both biotic and abiotic stresses. They are successfully responding to this challenge by developing crop varieties that are drought tolerant or that have a shorter growing cycle for early maturity. However, another challenge will be the adoption of those new varieties by farmers. Major research organizations involved in climate-smart agriculture are members of the CGIAR.

Interview

Dr. Jean Donné Rasasolofoniaina is a water and resource management specialist in Madagascar.

Can you describe your work related to climate change?

I do research in the field of water, especially in integrated water resources management in the face of climate change. I take local action to conserve water resources, especially mountain springs that are used in rural drinking water projects. My job is to educate local people on the conservation of watersheds from these sources and on the use of water to avoid waste. Awareness raising aims to eradicate bush fires that not only emit CO_2 but are the main threat to the sustainability of these sources. These awareness-raising actions are followed by reforestation activities and adapted cultivation practices. Theoretical research is also under way to establish a predictive model for the rational exploitation of these water resources.

What concerns you the most about climate change in your country?

Madagascar is ranked the fifth country in the world most vulnerable to climate change, according to the 2012 Maplecroft ranking, after Haiti, Bangladesh, Zimbabwe and Sierra Leone. Several regions of Madagascar are now experiencing the consequences of climate change, to varying degrees and vulnerabilities. Climate change has a huge impact on biodiversity, development and people's living conditions. Extreme weather events (cyclone, drought, floods) are becoming more and more periodic and intense. Extreme temperatures increase, especially minimum temperatures. Rainfall is becoming more intense and the rainy season is shorter, leading to significant consequences for rice production, the main food for the Malagasy people. It is these problems of water resource management that worry me about climate change.

What concerns you the most about climate change globally?

I am afraid for the future of mankind, especially in the face of the behavior of wealthy countries with respect to the conventions and actions undertaken on a global level. The reduction of CO_2 emissions seems to me to be utopian in the face of a desire to further develop the world economy. How far can we limit the use of traditional energies?

Can you describe a specific sustainability project in your area or country that is contributing to a "greener world" and lower carbon footprint?

The use of renewable energy is beginning to develop in my country, especially in rural areas. We see photovoltaic panels everywhere in the countryside of Madagascar, thanks to the funding of non-governmental organizations. Many small soil-conservation and reforestation projects exist in several regions, accompanied by awareness campaigns to reduce the practice of bush fires.

What major new steps could your government take to mitigate the growing threats from climate change?

The intervention of the Malagasy government is evident through the ratification of numerous international conventions such as COP21, where Madagascar has presented an ambitious 14% plan to reduce GHG emissions by 2020.

What is your reaction to the UNIPCC's accepted calculation by Prof. Myers of a minimum of 200 million climate change refugees by 2050 and the world's readiness to cope with this humanitarian and environmental challenge?

It is scary to see that figure of 200 million refugees, especially when there will be countries that will disappear because of sea level rise. I am thinking of the many small islands of the Pacific Ocean and the Indian Ocean. It is time that everyone responds in the right way to save our planet.

Can you describe a particular area of climate change that you are concentrating upon with research, or a grassroots development project?

As my field of research is in the water sector, I concentrate mainly on improving the management of this blue gold. Sometimes I feel very small in view of the magnitude of what must be done to save our planet but I think that each individual should take his share of responsibility.

Notes

1. IPCC, 2014.
2. Neumann, B., Vafeidis, A. T., Zimmermann, J. and Nicholls, R. J. (2015) Future Coastal Population Growth and Exposure to Sea-Level Rise and Coastal Flooding—A Global Assessment. PLoS ONE 10(3): e0118571. pmid:25760037).
3. IPCC (2014). Niang, I., O. C. Ruppel, M. A. Abdrabo, A. Essel, C. Lennard, J. Padgham and P. Urquhart (2014). Africa. In: Climate Change 2014: Impacts, Adaptation and Vulnerability. Part B: Regional Aspects, Chapter 22 (Africa). Contribution of Working Group II to the Fifth Assessment Report of the Intergovernmental Panel on Climate Change. pp. 1199–1265.
4. EJF (2009) No Place Like Home—Where next for climate refugees? Environmental Justice Foundation: London.
5. CRED, 2016 (generated from EM-DAT table on disaster subgroups, Kenya).
6. "Climate: Observations, projections and impacts, Kenya" Met Office, United Kingdom, www.metoffice.gov.uk, 2011.
7. CRED, 2016.

8. Mateshe, 2011; Government of Kenya, 2010.
9. CRED, 2016.
10. Anyah, R.O. and Semazzi, F.H.M. (2006).Variability of East African rainfall based on multiyear REGCM3 simulations. International Journal of Climatology, 27, 357–371.
11. Ruth Butterfield, SEI Oxford Centre, we Adapt, 'Climate changes in East Africa' March 30, 2011.
12. Githeko, A. K., Ototo, E. N. & Guiyun, Y. (2012) Progress towards understanding the ecology and epidemiology of malaria in the western Kenya highlands: Opportunities and challenges for control under climate change risk. Acta Tropica, 121, 19–25.
13. Maes, P., Anthony D. Harries, Rafael Van den Bergh, Abdisalan Noor, Robert W. Snow, Katherine Tayler-Smith, Sven Gudmund Hinderaker, Rony Zachariah and Allan, R. (2014) Can Timely Vector.
14. IPCC (2014). Niang, I., O.C. Ruppel, M.A. Abdrabo, A. Essel, C. Lennard, J. Padgham, and P. Urquhart (2014). Africa. In: Climate Change 2014: Impacts, Adaptation and Vulnerability. Part B: Regional Aspects, Chapter 22 (Africa). Contribution of Working Group II to the Fifth Assessment Report of the Intergovernmental Panel on Climate Change. pp. 1199–1265.
15. "Climate: Observations, projections and impacts, Kenya" Met Office, United Kingdom, www.metoffice.gov.uk 2011.
16. Jarvis, A., Ramirez-Villegas, J., Herrera Campo, B. & Navarro-Racines, C. (2012) Is Cassava the Answer to African Climate Change Adaptation? Tropical Plant Biology, 5, 929.
17. Thornton, P.K., Jones, P.G., Alagarswamy, G. and Andresen, J. (2009) Spatial variation of crop yield response to climate change in East Africa. Global Environmental Change, 19, 54–65.
18. Ramirez, J., A. Jarvis, I. Van den Bergh, C. Staver & Turner, D. (2011) Changing Climates: Effects on Growing Conditions for Banana and Plantain (Musa spp.) and Possible Responses. In: Yadav, S.S., Redden, R., Hatfield, J.L., Lotze-Campen, H. & Hall, A.J.W. (eds.) Crop Adaptation to Climate Change. Oxford, UK Wiley-Blackwell.
19. Scholtz, M., McManus, C., Leeuw, K., Louvandini, H., Seixas, L., Melo, C., Theunissen, A. and Neser, F. (2013) The effect of global warming on beef production in developing countries of the southern hemisphere. Natural Science, 5, 106–119.

20. Ifejika Speranza, C. (2010) Drought Coping and Adaptation Strategies: Understanding Adaptations to Climate Change in Agro-pastoral Livestock Production in Makueni District, Kenya. Eur. J. Dev. Res., 22, 623–642.
21. CRED, 2016.
22. IDMC, 2014c:4.
23. E. Araia, Refugiados ambientais—as primeiras vítimas do aquecimento global, Revista Planeta, 443, 36 (2009).
24. World Bank, Report No: AUS8099, Republic of Kenya, Kenya Urbanization Review. February 2016.
25. Awuor, C.B., Orindi, V.A. and Adwera, A.O. (2008). Climate change and coastal cities: the case of Mombasa, Kenya. Environment and Urbanization, 20(1), 231–242.
26. IPCC Fourth Assessment Report (2007), Working Group 1: The Physical Science Basis, Chapter 11, Figure 11.1.
27. "A Perspective on Sea Level Rise and Coastal Storm Surge from Southern and Eastern Africa: A Case Study Near Durban, South Africa" Andrew A. Mather and Derek D. Stretch, 2012. www.mdpi.com/journal/water.
28. Brown, S., Kebede, A.S. and Nicholls, R.J. (2011). Sea-Level Rise and Impacts in Africa, 2000 to 2100. University of Southampton, UK, 215 pp.
29. Ibid.
30. IPCC AR5.
31. DEA (Department of Environmental Affairs). 2013. Long-Term Adaptation Scenarios Flagship Research Programme (LTAS) for South Africa. Climate Trends and Scenarios for South Africa. Pretoria, South Africa.
32. Ibid.
33. Environmental Affairs Department: Republic of South Africa, "National Climate Change Response White Paper," 2011.
34. Ibid.
35. Environmental Affairs Department: Republic of South Africa, "National Climate Change Response White Paper," 2011.
36. "The impact of the current drought on the South African economy," Paul Makube, Senior Agricultural Economist at FNB, 2011.
37. Ibid.
38. Environmental Affairs Department: Republic of South Africa, "National Climate Change Response White Paper," 2011.

39. Mumura, Nobuo. "Sea-Level Rise Caused by Climate Change and Its Implications for Society." Editor, Kiyoshi Horikawa. Proceedings of the Japan Academy. Series B, Physical and Biological Sciences 89.7 (2013): 281–301. PMC. Web. 24 Apr. 2017.
40. Liz Heimann, "Climate Change and Natural Disasters Displace Millions, Affect Migration Flows" Issue # 7, Migration Policy Institute, MPI, December 10, 2015.
41. UNFCCC, "Botswana Intended Nationally Determined Contribution" http://www4.unfccc.int/ndcregistry/PublishedDocuments/Botswana%20First/BOTSWANA.pdf.
42. "Emergency Plan of Action (EPoA) Botswana: Floods" International Federation of Red Cross and Red Crescent Societies, March 11, 2017.
43. IPCC AR5.
44. 'Climate Change and Variability in Semiarid Palapye, Eastern Botswana: An Assessment from Smallholder Farmers' Perspective' Felicia O. Akinyemi, American Meteorological Society (AMS) 2017.
45. Working Group 1 (WG1) of the 5th Assessment Report (AR5) by the Intergovernmental Panel on Climate Change (IPCC).
46. African Development Bank, The Government of Botswana, Department of Agriculture and Agro-Industry, "Wastewater Reuse and Water Harvesting for Irrigation Study—Request for Technical Assistance Fund for Middle Income Countries" September 2011.
47. UNFCCC, "Botswana Intended Nationally Determined Contribution" http://www4.unfccc.int/ndcregistry/PublishedDocuments/Botswana%20First/BOTSWANA.pdf.
48. Guha-Sapir et al., 2004; Sida, 2008, Sida Arbetspapper: PD2 Botswana 2008-03-26.
49. "Wealth Accounting and the Valuation of Ecosystem Services" (WAVES) Country Report 2016 Botswana, June 2016.
50. Climate and Health Country Profile—2015, Botswana. World Health Organization, UNFCCC, 2015.
51. Ibid.
52. Botswana INDC, 2015.
53. Liz Heimann, Issue #7: "Climate Change and Natural Disasters Displace Millions, Affect Migration Flows" Migration Policy Institute, MPI, December 10, 2015.
54. William G. Moseley, "Agriculture on the Brink: Climate Change, Labor and Smallholder Farming in Botswana" MDPI, 27 June 2016.

7

India

"Without the clean flow of the river you can't ensure the human right."
Rajendra Singh, winner of the 2015 Stockholm Water Prize

India is the largest democracy and most populous nation in the world at 1.4 billion, and is projected to reach a population of 1.6 billion by 2050. It is the seventh largest country in the world by land mass at 3,287,263 km^2 (1,269,219 square miles) with a coastline of 7517 km (4700 miles). India has the seventh largest global economy with a gross domestic product GDP of 2.2 trillion (2016) yet ranks 130th in the HDI. It produces 28.2% of global CO_2 emissions. A deeply entrenched carbon economy renders substantial reduction in emissions an on-going challenge for many decades to come. India currently faces multiple climate change challenges that will intensify with devastating effect in the twenty-first century. These challenges include severe drought; coastal flooding; saltwater intrusion; water scarcity due in part to rapidly declining water tables from 4500 m in 1950 to 1500 m in 2017; food insecurity; skyrocketing food prices that will be unsustainable for tens of millions of citizens; significant and fatal health crises related to climate change; climate refugee flows in the millions; damage and destruction of infrastructure such as railways, roads, bridges and buildings; and extreme heat events. These multiple and sustained challenges will render climate change mitigation and adaption programs difficult to support effectively across the country.

As with India, the South Asian region will experience some of the harshest effects of climate change. Bangladesh will be particularly hard hit by sea level rise, flooding and staggering climate refugee numbers in the millions over the next century. Indeed, Bangladesh is rated as one of the five countries globally that is most impacted by climate change. The countries of the region, with a population of 1.8 billion, include Afghanistan, Bangladesh, Bhutan, Maldives, Nepal, India, Pakistan, and Sri Lanka. The land area is approximately 5.1 million km² (1.9 million square miles), which comprises 3.4% of the total global land surface.

Flooding and Drought

Flooding and drought have been perennial environmental problems throughout Indian recorded history. Yet since the 1950s there is a noticeable shift in the country toward more intense heat, deeper drought episodes, flooding and sea level rise induced flooding. India is profoundly dependent upon the annual monsoon rains to combat excess heat and provide vital irrigation as well as overall water sustenance. The southwest monsoons create heavy rainfall across most of India and the northeast monsoons cause substantial rainfall primarily in the southeastern coast region of the country. Thus erratic and interrupted monsoon patterns can have a devastating effect on the population and economic conditions. Two accurate climate change projections for India are increased drought episodes and higher flood episodes particularly in coastal areas besieged by sea level rise.

> India is the most flood distressed state in the world after Bangladesh, accounting for 1/5th of the global deaths every year with 30 million people displaced from their homes yearly. Approximately 40 million hectares of the land is vulnerable to floods, with 8 million hectares affected by it. Unprecedented floods take place every year at one place or the other, with

the most vulnerable states of India being Uttar Pradesh, Bihar, Assam, West Bengal, Gujarat, Orissa, Andhra Pradesh, Madhya Pradesh, Maharashtra, Punjab and Jammu & Kashmir.[1]

Moreover, there is an increase in cyclone incidents in the country due to climate change, which is a contributing factor to elevated flood risks. In recent years the Indian population has been devastated by a number of destructive flood crises. The 1987 Bihar flood impacted 29 million people; in 2008 the Bihar floods struck 2 million people; and in 2005 the Maharashtra flood caused 5000 fatalities. In 2011, a surge of intense rainfall triggered flood waters in the northern and eastern regions of the country that affected 10 million people, and in 2012 devastating floods in Assam state displaced 900,000 people. Severe flood episodes were also experienced in Uttarakhand in 2013, Srinagar in 2014 and Chennai in 2015.

With a coastline of 7517 km (4700 miles) India is clearly vulnerable to the inevitable outcome of pronounced sea level rise and coastal flooding. Three of the world's 20 most vulnerable cities to coastal flooding are in India: Chennai, Kolkata and Mumbai. Flooding will cause a range of traumatic events including loss of life, injury, property damage, infrastructure damage, damage to sewage and drainage systems, the spread of disease, damage to farmland and crops from saltwater intrusion and severe economic disruption. "Thousands of people have been displaced by ongoing sea level rises that have submerged low-lying islands in the Sundarbans." A 1 m sea level rise is projected to displace approximately 7.1 million people in India and about 5764 km^2 of land area will be lost, along with 4200 km of road. Around seven million people are projected to be displaced due to submersion of parts of Mumbai and Chennai if global temperatures were to rise by a mere 2 °C. The effects of global warming have also caused damage to coastal infrastructure, aquaculture and coastal tourism.[2] In the densely populated Ganges, Mekong and Nile River deltas, a sea level rise of 1 m could affect 23.5 million people and reduce the land currently under intensive agriculture by at least 1.5 million ha. A sea level rise of 2 m would impact an additional 10.8 million people and render at least 969,000 ha of agricultural land unproductive.

The other overwhelming threat to India is drought, which would be exacerbated by environmental change. In 2017, 256 districts experienced drought conditions. To complicate matters further, water tables have fallen from 4500 m in the 1950s to a dangerously low level of 1500 m in 2017. Water insecurity, inefficient irrigation systems and the overuse of water resources are having devastating impacts that will worsen as drought conditions persist. Currently, the government of India is planning a USD 165 billion water diversion project for drought-prone regions. In sum, 15,000 km of artificial waterways will be connected to 37 rivers. Approximately 68% of the agriculture land in the country is vulnerable to drought. Rising temperatures, extreme heat and severe water scarcity are projected with certainty in the twenty-first century.

Extreme Weather

Extreme weather events are a common and tragic experience in India. The country has 22 states that are disaster prone. Drought effects 68% of the country, cyclones reach 8% of the country and 12% is subject to flooding. All of these extreme weather patterns are anticipated to worsen over the coming decades. In a country of 1.4 billion people, these impacts affect millions of citizens each year. According to recent studies, "if the process of global warming continues to increase, resulting climatic disasters would cause a decrease in India's GDP to decline by about 9%, with a decrease by 40% of the production of the major crops." A temperature increase of 2 °C in India is projected to displace seven million people, with a submersion of the major cities of India like Mumbai and Chennai.[3] An analysis of environmental change in India observed that,

> Extreme weather events such as severe storms, floods, and drought have claimed thousands of lives during the last few years and have adversely affected the lives of millions and cost significantly in terms of economic losses and damage to property. India, like other developing countries, is poorly equipped to deal with weather extremes. Hence, the number of people killed, injured, or made homeless by natural disasters has been increasing rapidly.[4]

Health, Human Development and Food Security

India has a high vulnerability scenario due to the fact that climate change has more severe and disproportionate effects upon the poor. India has one of the lowest HDI rankings in the world and a population of 224 million living below the international USD 1.90 per day poverty line. Thus climate change will pose particular hardships upon close to 20% of the population. A recent report noted some of the serious environmental consequences facing India.

> Nearly 700 million of her over one billion population living in rural areas directly depends on climate-sensitive sectors (agriculture, forests, and fisheries) and natural resources (such as water, biodiversity, mangroves, coastal zones, grasslands) for their subsistence and livelihoods. Heat wave, floods (land and coastal), and draughts occur commonly. Malaria, malnutrition, and diarrhea are major public health problems. Any further increase, as projected in weather-related disasters and related health effects, may cripple the already inadequate public health infrastructure in the country.[5]

Specific health threats are expected to increase in India as the adverse effects of climate change become more pronounced. Segments of the population, such as the impoverished, elderly and disabled, will face the additional burden of coping with environmental threats. "Globally, temperature increases of 2–3 °C would increase the burden of a number of people who, in climatic terms, are at a risk of malaria by around 3–5%, i.e. several hundred million." Further, the seasonal duration of malaria would increase in many currently endemic areas. In India, the transmission windows for malaria are predicted to increase with climate change from 4–6 months to 7–9 months in a year in Jammu, Kashmir and Madhya Pradesh and from 7–9 months to 10–12 months in Uttar Pradesh.[6] Both floods and droughts increase the risk of diarrheal diseases. Major causes of diarrhea linked to heavy rainfall and contaminated water supplies are cholera, cryptosporidium, *E. coli* infection, giardia, shigella, typhoid and viruses such as hepatitis A.[7] In 2030, the estimated risk of diarrhea will be up to 10% higher in some regions than if no climate change occurred.[8]

The Human Security paradigm, first articulated in a landmark 1994 UNDP report, notes seven threats to individual security and development. One of these threats is food insecurity. India is facing dramatic and long-term threats to food security due to the current and escalating problems of climate change upon the agriculture sector, which will profoundly effect human life and health. Moreover, food prices are anticipated to increase sharply thus imposing severe hardships on the millions of Indians who live below the international poverty line. Climate change threats confront the agriculture industry and national food supply.

Studies by the Indira Gandhi Institute portray a difficult scenario vis-à-vis agriculture production and supply:

> In South Asia, climate change leads to a 14 per cent decline in rice production relative to no climate change scenario, a 44 to 49 percent decline in wheat production, a 9 to 19 decline in maize production and 12.2 to 19.6 percent decline in sorghum production. The results are mixed for the East Asia and Pacific sub-region. Rice production declines by about 10 percent while wheat production increases slightly. Between 2000 and 2050, even with no climate change, the price of rice would rise by 62 percent, maize by 63 percent, soybean by 72 percent, and wheat by 39 percent. The climate change results in additional price increases—a total of 32 to 37 percent for rice, 52 to 55 percent for maize, 94 to 111 percent for wheat, and 11 to 14 percent for soybeans. Similarly, the meat prices are 33 percent higher by 2050 without climate change and 60 percent higher with climate change and no CO_2 fertilization of crops.[9]

Studies by the Indian Agricultural Research Institute indicate the possibility of a loss of between 4 and 5 million tons in wheat production in the future with every temperature rise of 1 °C throughout the growing period, whereas rice production is slated to decrease by almost a ton/hectare if the temperature rises by 2 °C. In Rajasthan a 2 °C rise in temperature was estimated to reduce production of pearl millet by 10–15%.[10] A report on the Indian agriculture scenario in the context of severe environmental dislocation also signaled a problematic future for the country. As the report noted,

> food security is both directly and indirectly linked with climate change. Any alteration in the climatic parameters such as temperature and humidity

which govern crop growth will have a direct impact on quantity of food produced. Indirect linkage pertains to catastrophic events such as floods and droughts which are projected to multiply as a consequence of climate change leading to huge crop loss and leaving large patches of arable land unfit for cultivation which hence threatens food security.[11]

It has been projected that under the scenario of a 2.5–4.9 °C temperature rise, rice yields will drop by 32–40% and wheat yields by 41–52%. This would cause GDP to fall by 1.8–3.4%.[12]

As noted in the HDI ranking for India, serious health and social indicators are in need of considerable progress. The Dalit population, for example, numbering 160 million, is struggling under the dual burdens of historic discrimination and extreme poverty. According to a recent report addressing food security and health issues,

> simulations using dynamic crop models indicate a decrease in the yield of crops as temperature increases in different parts of India. This is likely to threaten the food security in the country where malnutrition is already an important public health problem. Malnutrition causes millions of deaths each year, from both a lack of sufficient nutrients to sustain life and a resulting vulnerability to infectious diseases such as malaria, diarrhea, and respiratory illnesses. In India, almost half of the children under age five and more than one-third of the adults are undernourished.[13]

Yet all Indian citizens who are marginalized and impoverished as well as those citizens living in vulnerable regions such as Chennai or Mumbai, will be forced to confront significant environmental threats in the twenty-first century that will impact health, agriculture, food security and economic opportunities. Climate change will magnify many existing health and socio-economic problems facing the Indian population.

Climate Change Refugees

India will face dramatic climate change migration flows both internally and to some extent externally. The impacts will be far worse than most other nations in the world. "A one meter sea level rise is projected to displace approximately 7.1 million people in India and about 5764 km^2 of land area will be lost, along with 4200 km of road. Around seven million people are projected to be displaced due to submersion of parts of Mumbai and Chennai if global temperatures were to rise by a mere 2 °C."[14] A 2007 study using satellite data identified coastal areas at less than 30 feet above sea level with a substantial flood risk. The conclusions were that 634 million people inhabit these low-lying areas. This represents 10% of the world's population living in 2% of the world's land mass. "The study also found the 10 countries with the largest share of their populations in low-elevation coastal zones are Bangladesh, China, Egypt, Gambia, India, Indonesia, Japan, the Philippines, Thailand and the United States."[15] A 2009 report by the International Organization for Migration in collaboration with the United Nations University and the Climate Change, Environment and Migration Alliance projected climate refugee flows in numbers ranging from "200 million to 1 billion migrants from climate change alone, by 2050" and noting that the "environmental drivers of migration are often coupled with economic, social and developmental factors that can accelerate and to a certain extent mask the impact of climate change."[16]

A recent example of large-scale mass displacement triggered by environmental crisis is the devastating Gujarat floods in 2005 that led to the displacement of 250,000 people. Several coastal cities face significant flood risk thereby heightening the forced migration of citizens in mass numbers. India has nine coastal states and territories including Andhra Pradesh, Goa, Gujarat, Maharashtra, Karnataka, Kerala, Maharashtra, Odisha, Tamil Nadu, West Bengal, Daman and Diu, and Puducherry. The combined population is approximately 575 million. All of these regions and main cities, such as Chennai, Kolkata and Mumbai, will experience strong forced migration pressures from flooding and environmental hazards.

City Profile: Mumbai

Mumbai has a population of 21 million (2017) and is the fourth most populated city in the world and the largest city in India. It is located on the west coast of India and is a vital sea port. The total land area is 437.71 km^2 (169 square miles). The larger Mumbai Metropolitan Region covers a sprawling area of 4355 km² (1681 square miles). The main climate change related threats to the city residents include sea level rise, flooding, saltwater intrusion, infrastructure damage, health related problems including the spread of disease, damage to drainage and sewage systems and the disruption of vital services such as transportation. An estimated 41 million citizens in India live in slums and 750,000 to 1 million in the Dharavi slum, which is the third largest in the world. These citizens living in challenging environments with low income levels will be heavily burdened by the additional effects of climate change.

Mumbai has an average elevation of only 14 m (46 feet). A study of 136 port cities showed that the population exposed to flooding linked with a 1-in-100-year event is likely to rise dramatically from 40 million currently to 150 million by 2070.[17] Mumbai as a port city is included in this high risk category. "The value of global assets exposed to flooding is estimated to rise to USD 35 trillion, up from USD 3 trillion today."[18]

Innovation and Adaption Strategies

In 2016 India ratified the Paris Agreement.
Government goals:

1. to increase the share of non-fossil-based power generation capacity to 40% of installed electric power capacity by 2030;
2. by 2020, to reduce the emissions intensity of GDP by 20% to 25% below 2005 levels;
3. to increase the solar ambition of the government's National Solar Mission to 100 GW installed capacity by 2022;
4. to have solar power in every home by 2019 and invest in 25 solar parks;
5. to raise wind energy production from 50,000 to 60,000 MW by 2022 (the 12th Five Year Plan—National Wind Energy Mission);
6. Prime Minister Modi announced a goal of 100 sustainable "smart cities";
7. to provide electricity to 25% of the population by 2022;
8. in 2008 Prime Minister Singh released India's first National Action Plan on Climate Change (NAPCC).

Interview

Ashok Kumar Rajagopal is a dedicated climate change and water resources specialist based in Chennai, India, and has conducted extensive research in the areas of water resources and coastal engineering. He has master's degrees in marine geology, petrology and mineralogy and ocean engineering from the University of Madras and from the Indian Institute of Technology, Madras.

Can you briefly describe your work related to climate change?

I am a water resources specialist with a main focus on climate change and its relationship with energy and unprecedented events that are taking place on this planet. I monitor and see the changes that are taking place due to due to human induced (or anthropogenic) events after the post-industrial era (after 1850).

What concerns you the most about climate change in your country?

India is a tropical country with an estimated population of 1,267,401,849 as of July 1, 2014. India's population is equivalent to 17.5% of the total world population. It ranks second in the list of countries by population. The population density is 386 people km^{-2}. The country has a total land area of 3,287,263 km^2 with a coastline of 7516.6 km in length and 29 states and seven union territories. India's GDP is around USD 2073. Its per capita electricity consumption stands at between 900 and 950 kWh.

What concerns you the most about climate change globally?

Globally, every country advocates for low carbon emissions but simultaneously fossil fuels are being extracted or produced in huge quantities and the cost of fossil fuel has come down drastically. This makes all of the countries, right from the least developed to developed, use fossil fuels. This addiction to fossil fuels results in more emissions of CO_2 into the atmosphere, which is causing unprecedented events such as heavy rainfall, cloud bursts, intense and unpredicted development of storms and cyclones, melting of glaciers and ice caps in high altitude regions, droughts and flooding of coastal cities and sea level rise.

Can you describe a specific sustainability project in your area or country that is contributing to a "greener world" and lower carbon footprint?

A project started in July 2015 and runs to June 2019. Renewable Energy Technology Packages for Rural Livelihoods (RETPRLs) is operating in three selected states—Assam, Odisha and Madhya Pradesh. The RETPRLs include: solar lighting systems; solar and/or biomass waste-powered micro grids for common facilities; solar irrigation pumps; improved commercial biomass cooking stoves; poultry-litter-based biogas plants; poultry-litter-based briquetting units; solar dryers for vegetables, spices and fish; solar-powered milk chillers; and cold rooms for storage of horticultural produce. The target livelihood sectors include poultry, fisheries, dairy, horticulture, khadi (homespun cloth) and silk weaving, bamboo and commercial cooking (e.g., tea stalls, sweet shops and street food vendors).

Under the guidance of and finance from the UNDP (USD 800,000), along with the Global Environmental Facility (USD 4,006,849), government of India (USD 10,000,000) and the private sector (USD 8,233,767), the project is being carried out in the states of Assam, Madhya Pradesh, and Odisha.

What steps could your government take to mitigate the growing threats from climate change?

The government of India is taking very good measures described in documents submitted to the UNFCCC. See: http://www4.unfccc.int/Submissions/INDC/Published%20Documents/India/1/INDIA%20INDC%20TO%20UNFCCC.pdf.

What is your reaction to the UNIPCC's accepted calculation by Prof. Myers of a minimum of 200 million climate change refugees by 2050 and the world's readiness to cope with this humanitarian and environmental challenge?

All developed and developing nations have to accept climate change refugees and accommodate them to cope with this environmental challenge. Climate change events are caused not by one country, it is the cumulative effect of all, so we all have to accept climate change refugees.

Can you describe a particular area of climate change that you are concentrating upon with research or a grassroots development project.

The main cause of climate change is energy. We have to balance the input and output of energy that is being received from the Sun.

The natural hydrological cycle is disrupted by damming all surface water to harness hydropower and not leaving enough water to go downstream and to the oceans. This has a link with the development of desalination plants all along the coastlines of the world. These desalination plants consume lots of energy and cause severe disruption to the natural hydrological cycle, which is the main cause of unprecedented events such as excess rain and drought. Living in uninhabitable places and using fossil fuels has caused climate change. We are consuming more energy to live in "uninhabitable" locations like hot deserts and extreme cold.

Interview

Dr. Allen Swagoto Baroi is a medical practitioner and public health professional. He completed a master's in public health, concentrating on global health at Thammasat University, Thailand, and a bachelor's in medicine and surgery from the University of Dhaka, Bangladesh. He is passionate about working with underprivileged people in challenging environments and patients suffering from disease related stigma.

Can you describe your work related to climate change?

Global climate change is one of the greatest upcoming threats that has been changing at an unpredictable rate in South Asia for the past few decades. I had the opportunity to work as a physician with an INGO dedicated to the elimination of topical neglected diseases. In the tenure of my service, I dealt with patients effected by several pathological skin conditions directly or indirectly related to climate change. I saw a number of clinical conditions, including leprosy, squamous cell carcinomas and vector-borne infectious diseases.

Leprosy is one of the oldest known diseases with a low prevalence and incidence rate. Although the number of cases has been declining in recent years, a significant number of the population in tropical countries are still affected by the disease. Tropical countries like India, Bangladesh, Brazil and Indonesia are recognized worldwide for their vulnerability to global warming and climate change. Climate change has a negative influence on the disease. Elimination programs have not been successful due to the continual change in climate and there are cases that go undiagnosed and untreated.

In recent times patients have been identified who are suffering from squamous cell carcinomas that are a secondary development of other premalignant disease conditions. A literature review reveals that climate change has a direct effect on this particular skin condition. Increased ultraviolet radiation induces carcinogenesis as a result of depletion of the ozone layer in atmosphere. The chance of human ultraviolet exposure has been greater than before due to increased temperatures in tropical regions. This leads to an enhanced incidence rate of different types of pathological conditions.

Certain vector-borne diseases are likely to expand in South Asia due to climate change. Changes in humidity, temperature and increased adoptability affects the biological and ecological system. Several outbreaks of certain infectious vector-borne diseases like dengue, malaria and scabies have been observed in the past few decades in this region.

What concerns you the most about climate change in your country?

Bangladesh is one of the South Asian countries recognized worldwide for its vulnerability to climate change and global warming. Several sudden severe and catastrophic events like floods, landslides, cold spells, droughts, cyclones and tornedos have ravaged the country in the past ten years. A rise in sea level of 0.5 m over the last 100 years has already eroded 65% of the land mass of several islands in the south of Bangladesh in the Bay of Bengal.

The effects of climate change become more devastating for a developing country like Bangladesh because several determinants of health are interlinked, such as poor infrastructure and policy, education and culture.

The most important issue that concerns me is female education and empowerment. In most cases women and children are the vulnerable groups that are most affected by catastrophic events. In a patriarchal society like Bangladesh, women and children still require a decision from a father or husband to safeguard them during any disaster event.

What concerns you the most about climate change globally?

Climate change is a continuous process. Several issues are interlinked with global warming. The race for development and financial stability among nations enhances the process of global warming. The concerns about climate change are as follows:

1. insufficient funds;
2. ineffective policies;
3. ineffective monitoring systems;
4. lack of regional cooperation;
5. unplanned urbanization and centralization of public facilities;
6. deforestation;
7. lack of responsibilities.

Can you describe a specific sustainability project in your area or country that is contributing to a "greener world" and lower carbon footprint.

Bangladesh is a developing country in South Asia with a huge population. The country adopted various projects that contribute to a "greener world" and it has achieved remarkable progress through "The Bangladesh Climate Change Resilience Fund."

The main expected results and activities are as follows:

1. food security, social protection and health are improved;
2. disaster risk management is addressed comprehensively;
3. resilient infrastructure is built;
4. the knowledge base is increased;
5. carbon emissions are mitigated, and development is less dependent on carbon;
6. capacities are built and institutions strengthened.

Key achievements of the project include:

1. the establishment and operation of governance structures along with recruitment of technical human resources;
2. the construction of multipurpose cyclone centers and roads in affected areas;
3. adequate measures taken to reduce calamities in targeted areas, such as reducing forest degradation and increasing forest coverage through proper planning and monitoring in coastal and hilly areas;
4. a renewable energy development project promoting solar energy utilization that reduced dependency on energy from a coal power plant;
5. taking different measures on disease control, such as vector borne diseases that are directly impacted by climate change.

What steps could your government take to mitigate the growing threats from climate change?

Bangladesh is one of the low-income countries in Southeast Asia. Its vulnerability to climate change is of great concern due to the low level of development and factors like geographical position and rapid change of climate. More people are now exposed to risk, and opportunities for internal migration are limited. The government has already taken measures to mitigate the growing threat of climate change but it still needs to address the following areas for a sustainable progress:

1. decentralization of power for rapid response to disaster;
2. form a policy for emergency funds and rapid response;
3. gender equality and female empowerment;
4. corruption;
5. adequate resources to build or rebuild infrastructures;
6. to adopt ecofriendly policies such as banning vehicles responsible for air pollution, effective and planned tree plantation and a monitoring system for industries responsible for environmental pollution.

What is your reaction to the UNIPCC's accepted calculation by Prof. Myers of a minimum of 200 million climate change refugees by 2050 and the world's readiness to cope with this humanitarian and environmental challenge.

The calculation adopted by the UNIPCC is certainly a good initiative and an alarming situation for both developed and developing countries. There are challenges that face both refugees and host countries.

For refugees, the difficulties are:

1. adapting to new geographical area and culture;
2. meeting basic needs for living (food, water, education, health, shelter and so on);
3. reorganization by the host country;
4. and finding a source of income and security.

For the host country, the difficulties are:

1. policy analysis for refugees;
2. budget allocation for refugee resettlement;
3. dealing with cultural invasion and conflicts;
4. and combating adverse situations like international terrorism, illegal drug trade and a global black market.

This is why "prevention is better than cure" and sustainable measures should be taken to reduce climate changes as well as global warming to avoid such a situation by 2050.

Can you describe a particular area of climate change that you are concentrating upon with research or a grassroots development project?

I had been working as a medical practitioner with an international NGO dedicated to eliminating neglected tropical diseases. In the tenure of my service I experienced different areas of stigma related to a particular disease. The disease itself had a high prevalence and incidence rate in past years in this particular region due to climate change. Being a physician and public health professional, I am concentrating on the following areas that need to be considered when dealing with climate change:

1. gender inequality and vulnerability to disaster;
2. social stigma and climate change;
3. and universal primary education.

Interview

Dr. Md. Mizanur Rahman is a project director of Bangladesh Delta Plan 2100 Formulation Project, Ministry of Planning (superintending engineer with BWDB, MoWR). He has 28 years of experience in research, academia and consultancy, and field experiences in hydrology, water resources engineering, flood risk management, climate change impact adaption, irrigation and hydraulic structures construction. Dr. Md. Mizanur Rahman is adunct faculty with the Institute of Disaster Management and Vulnerabilities Studies, University of Dhaka. He is a member of the Research and Academic Committee (RAC) of the Institute of Water and Flood Management, BUET. He was a post doctorate fellow in Nipissing University, Canada, and achieved a best article award from the Journal of Hydrologic Engineering, ASCE Publishers (awarded on "World Environment & Water Resources Congress 2014" at Portland,

Oregon, USA). He also achieved the Netherlands Fellowship and TCS Colombo Plan awards. His name was placed on the approved list of visiting post doctorate fellowship program in NSERC, Canada. He has published numerous articles in international peer-reviewed journals and conferences and he is interested in further studies in climate related research.

Can you describe your work related to climate change?

Bangladesh is a tropical monsoon climate country. With enormous crisscross river systems, having mostly low lying deltaic landforms, receiving maximum run-off from the Hindu Kush Himalayan region and being in close proximity to the Bay of Bengal, the country has one of the largest deltas in the world and is one of the most vulnerable to climate related disasters. Bangladesh is already experiencing increased temperature, changes in rainfall pattern and distribution, sea level rise and salinity intrusion at an accelerated rate and increased disaster intensity, which will become greater issues in the future.

Looking ahead, climate change will also become a major threat to Bangladesh's aspirations to ensure food security, sustainable development and poverty alleviation. Agriculture, including crops and horticulture, forestry, livestock and fisheries, are the most climate-sensitive sectors. These sectors must therefore adapt to the impact of climate change to improve the resilience of food production systems in order to feed a growing population. Since water is critical to agriculture and livelihoods, managing the water resources and addressing the vagaries of climate change will have to be co-integrated national strategies. That remains a major long-term challenge to the Bangladesh economy, which must be addressed as an integral part of the overall development agenda and under the "Bangladesh Delta Plan 2100" formulation project. The Bangladesh Delta Plan 2100 formulation project will develop strategies that contribute to disaster risk reduction, climate change resilience and adaption, water safety, food security, environmental safety and the economic development of the country. It is being implemented with technical and financial assistance from the government of the Netherlands, the World Bank and IFC 2030. The government of Bangladesh appointed me as project director for this project.

What concerns you the most about climate change in your country?

Bangladesh is one of the top ten nations that are most vulnerable to climate change concerns of sea level rise, increasing flood inundation, scarcity of water during the dry season, drainage congestion and water logging, river erosion, migration and ground water depletion. Due to climate change over the next 100 years, there are a number of predicted consequences.

An accelerated sea level rise (ASLR) of 2 mm/year will increase mean sea level by 1 m or more by 2100, which will increase tidal flooding. The consequences of this will be flood inundation of up to 20% (up to 30,000 km^2) of the total land area. This sea level rise will increase salinity in additional land areas, result in loss of farm land and livelihoods, out migration, destruction of infrastructure and destruction of mangroves in the Sundarbans, a UNESCO world heritage site.

The predicted rise in sea level may inundate up to 75% of the Sundarbans; highly specialized plants and animals will migrate to the north or disappear while less specialized organisms may be in a position to adapt in the newly created environment. Higher future CO_2 levels could benefit forests with fertile soils in the northeast. However, increased CO_2 may not be as effective in promoting growth in the west and southeast, where water is limited.

What concerns you the most about climate change globally?

In the moderate scenario posited by the IPCC, global temperatures are expected to rise by 2 °C by 2050 and by another 3–4 °C by the end of this century. In cooler regions of the world, it is possible that initially global warming may have some favorable impacts on crop yields, and declining yields will only kick in after warming exceeds 2 °C, beyond 2050.

Can you describe a specific sustainability project in your area or country that is contributing to a "greener world" and lower carbon footprint?

Bangladesh emissions of CO_2 are not a concern, so rather than mitigation CO_2 emissions, policies, acts and frameworks have been formulated for adaption to climate change impact. However, there are many projects for mitigation of environmental degradation like afforestation in coastal areas and revitalization of water bodies and wetlands in Bangladesh. The Clean Development Mechanism is one of the major projects in Bangladesh.

What steps could your government take to mitigate the growing threats from climate change?

In 2005, the government of Bangladesh formulated the National Adaptation Programme of Action after extensive consultations with stakeholders, professionals, scientists, academicians and other people of civil society. Continuing this process, the government has also adopted the Bangladesh Climate Change Strategy and Action Plan (BCCSAP), which will be the main basis of our efforts to combat climate change over the next ten years. The BCCSAP will remain as a "living document" to incorporate a future uncertain about the timing and exact magnitude of many of the likely impacts of climate change. The BCCSAP therefore, anticipates periodical revision, as required. Responsibility for implementing the various components of the BCCSAP will lie with line ministries and agencies, which will work in partnership with each other and with civil society and the business community.

What is your reaction to the UNIPCC's accepted calculation by Prof. Myers of a minimum of 200 million climate change refugees by 2050 and the world's readiness to cope with this humanitarian and environmental challenge.

The coastal zone of Bangladesh is most vulnerable to accelerated sea level rise, increasing salinity and water logging, tidal fluctuation and increasing intensity of cyclones and storm surges, which will ultimately lead to huge losses of farming land, homesteads and livelihoods, accompanied by severe health impacts (due to scarce fresh water and salinity in drinking water), which will accelerate the out-migration of people from the coastal zone into urban slums. The total area of the coastal zone is 47,203 km^2, which contains a population of 38.5 million. In Bangladesh, studies show that there are already some statistics indicating that people from this area are already starting to out-migrate towards urban slams and this will increase with a long-term trend in 2025 and 2050. So, it is clear that the calculation by Prof. Myers of a minimum 200 million climate change refugees by 2050 will be validated in Bangladesh.

Can you describe a particular area of climate change that you are concentrating upon with research or a grassroots development project?

Bangladesh is a unique hydrological land mass formed at the estuary of the Bay of Bengal and the Ganges–Brahmaputra–Meghna basin and is known as the largest delta in the world. This remains a major long-term challenge for the Bangladesh economy and must be addressed as an integral part of the overall development agenda under the Bangladesh Delta Plan 2100 formulation project.

Notes

1. Global Warming and its Impact on Climate of India (https://www.climateemergencyinstitute.com).
2. Ibid. 3. EJF (2009) "No Place Like Home—Where next for climate refugees?" Environmental Justice Foundation: London.
3. Global Warming and Its Impact on Climate Change of India (https://www.climateemergencyinstitute.com).
4. Majra, J.P., and A. Gur. "Climate Change and Health: Why Should India Be Concerned?" Indian Journal of Occupational and Environmental Medicine 13.1 (2009): 11–16. PMC.
5. Ibid.
6. Available from: http://www.nplindia.org/npl/climate change impacts on human health in India: Key sheet 9.htm.
7. Wilson M.L. Ecology and infectious disease. In: Aron J.L., Patz J.A., editors. Ecosystem change and public health: A global perspective. Baltimore: Johns Hopkins University Press; 2001. p. 283–324.
8. World Health Organization. Climate Change and Human Health: Risks and Responses. Summary. Geneva: WHO; 2003.
9. S. Mahendra Dev, "Climate Change, Rural Livelihoods and Agriculture (focus on Food Security) in Asia-Pacific Region" Indira Gandhi Institute of Development Research, Mumbai August 2011. http://www.igidr.ac.in/pdf/publication/WP-2011-014.pdf.
10. Abhimanyu Shrivastava, "Climate Change and Indian Agriculture," International Policy Digest, August 2016, and Indian Agricultural Research Institute.
11. Abhimanyu Shrivastava, "Climate Change and Indian Agriculture," International Policy Digest, 22 August 2016.
12. Indian Meteorological Department. Available from: http://www.imd.gov.in.
13. National Family Health Survey (NFHS-3), 2005–06. Volume 1. India: Mumbai: IIPS; 2007. International Institute for Population Sciences (IIPS) and Macro International.

14. "Global Warming and its Impact on Climate of India" https://www.climateemergencyinstitute.com.
15. "The rising tide: assessing the risks of climate change and human settlements in low elevation coastal zones" Gordon McGranahan, Deborah Balk and Brigit Anderson, in Environment and Urbanization, April, 2007.
16. "Migration, Climate Change and the Environment" Compendium of IOM.s Activities. International Organization for Migration, 2009.
17. Nicholls, R., et al. (2008), "Ranking Port Cities with High Exposure and Vulnerability to Climate Extremes: Exposure Estimates," *OECD Environment Working Papers*, No. 1, OECD Publishing, Paris. https://doi.org/10.1787/011766488208.
18. "Climate Risks and Adaptation in Asian Coastal Megacities" A Synthesis Report, World Bank, 2010.

8

The Middle East: Egypt, Israel, Jordan

> *"Trees are poems that Earth writes upon the sky."*
> Gibran

Egypt

Egypt is an ancient civilization with a profound influence upon Arab culture and politics. The country borders the Mediterranean Sea to the north and is home to the Nile River, the longest river in the world and controls the vital Suez Canal. The population is 94.7 million (2017), the land area is 1,001,449 km^2 (386,662 square miles) with a coastline of 2900 km (1800 miles) that comprises the Mediterranean Sea, the Gulf of Suez and the Gulf of Aqaba. Approximately 40% of the population is urban which includes the major metropolis of Cairo (19 million people) and Alexandra (5 million people) whose citizens face tremendous environmental challenges. The GDP is USD 346 billion (2016) and overall life expectancy is 71.4 years for men and 74.2 for women with a country HDI rank of 108. With the sprawling Sahara desert to the south, high temperatures, a dry climate and extensive coastline, Egyptians face considerable climate change threats including drought, sea level rise, coastal flooding, saltwater intrusion, desertification, sandstorms, heat waves, food insecurity and elevated health stress levels. In 2016, the country contributed 0.5% of global CO_2 emissions.

Flooding and Drought

The Nile river and delta are vital to the population, agriculture and commerce of Egypt and home to more than 80 million citizens. Yet climate change impacts of flooding and rising heat levels will dramatically affect the region and population. The extremely low sea level of the Nile River and delta at 1 m is cause for immediate concern by Egyptian authorities given the certainty of sea level rise over the coming decades. "The northern third of the delta is lowering at the rate of about 4–8 mm per year due to compaction of strata underlying the plain, seismic motion, and the lack of sufficient new sediment to re-nourish the delta margin being eroded by Mediterranean coastal currents. While the coastal delta margin is being lowered, sea level is also rising at a rate of about 3 mm per year. At present rates, saline intrusion is now reaching agricultural terrains in central delta sectors—the coastal 20 to 40 km of delta surface will be underwater by the end of this century."[1] Moreover, only 4% of the land mass of the country is inhabited, with extensive crowding in the major cities of Alexandria and Cairo and the Nile region. In addition, the population is expected to double by 2100 with land stress currently evident and water insecurity expected to reach critical levels by 2025. Water capacity per capita is expected to decline by 50% by 2050.

A report commissioned by the Arab Forum for Environment and Development observed three alarming scenarios with a high probability of occurring in the near term. According to the report: nearly 12 million Egyptians may also need a safe place to go if sea levels rise and drown the Nile Delta, whose dense river network provides water for the bulk of Egypt's farmland; in "extreme-case scenarios" of a 5 m rise in sea level rise, more than half the Nile Delta would experience devastating effects; and ten major cities, including Alexandria and Port Said, would be threatened.[2] An analysis of global coastal flooding observed that ten countries with the most people living in low elevation coastal zones

(LECZs) together account for about 463 million people and that the most vulnerable countries with 125 million people affected are Vietnam, Bangladesh and Egypt.[3] The Intergovernmental Panel on Climate Change (IPCC) has calculated that the Mediterranean will rise 30 cm to 1 m in the twenty-first century. A 2007 World Bank study estimated that a 1 m sea level rise could displace 10% of the population. "Egypt (26 million; 38% of its total population) and Nigeria (7.4 million; 5.9% of its total population) were the countries with the highest population in the LECZ in the African continent in 2000, ranking at places 6 and 7 globally, the Egyptian LECZ along the Mediterranean coast and the Nile delta (1075 people/km^2) was almost as densely populated as the LECZ of Japan (1250 people/km^2) or Bangladesh (1154 people/km^2) in 2000. By 2030, population density along the Egyptian coast is expected to increase to 1902 people/km^2 and to 2681 people/km^2 by 2060."[4]

All sectors of the country will be severely impacted by sea level rise and flooding and the human health costs will be substantial. Moreover, the country will be burdened by heavy infrastructure damage costs and essential mitigation efforts. "In one study that considered the impact of a 1 m SLR for 84 developing countries, Egypt was ranked the 2nd highest with respect to the coastal population affected, 3rd highest for coastal GDP affected and 5th highest for proportion of urban areas affected and around 15% (2.7 million people) of Egypt's coastal population could be affected by a 10% intensification of the current 1-in-100-year storm surge combined with a 1 m SLR."[5]

Another major concern confronting Egyptian authorities and the population is drought. As dry conditions already prevail and desert covers 262,800 square miles (680,650 km^2), which is approximately 65% of the country, the incidence of drought and indeed further desertification is projected to increase. Egypt receives about 80 mm of rainfall yearly and only 6% of the land is arable and suitable for agriculture. As water stress intensifies, the ability to irrigate ever more precious agriculture lands will become strained thus leading to further food insecurity, dramatically rising food prices and affected livelihoods. Egypt is a desert country that relies heavily upon the Nile River to supply 97% of the water supply. The population is projected to reach 130 million by 2030 at a time when drought, water insecurity and climate change adversity will be intensified.

Climate change will have a pronounced impact upon temperature and drought conditions. According to recent research, scenarios for North Africa and Egypt indicate, "an average rise in annual temperatures, higher than the average expected for the planet. Heat waves would then be more numerous, longer and more intense. North Africa would be particularly affected by droughts that would be more frequent, more intense and longer-lasting. The projections also announce a drop of 4–27% in annual rainfall. The water deficit will be worsened by increased evaporation and coastal aquifers will become more salty. The sea level could rise by 23–47 cm. by the end of the 21st century."[6]

Additional weather conditions facing the country include high winds and sandstorms that have severe consequences. In a dry, semi-arid country these climate factors intensify agriculture and water stress. The Egyptian Meteorological Authority has observed that a "phenomenon of Egypt's climate is the hot spring wind that blows across the country. The winds, known to Europeans as the sirocco and to Egyptians as the khamsin, usually arrive in April but occasionally occur in March and May. The winds form in small but vigorous low-pressure areas in the Isthmus of Suez and sweep across the northern coast of Africa. Un-obstructed by geographical features, the winds reach high velocities and carry great quantities of sand and dust from the deserts. These sandstorms, often accompanied by winds of up to 140 kilometers per hour, can cause temperatures to rise as much as 20 °C in two hours."[7]

Extreme Weather

The most reliable projections for Egypt in terms of extreme weather include rising temperatures, erratic rainfall, sandstorms and extreme heat. In general, North Africa and Egypt will experience rainfall reduction that in some regions will aggravate persistent drought conditions. "Over North Africa under the SRES A1B scenario, both annual minimum and maximum temperature are likely to increase in the future, with greater increase in minimum temperature. The faster increase in minimum temperature is consistent with greater warming at night, resulting in a decrease in the future extreme temperature range."[8]

Health, Human Development and Food Security

The Egyptian population is relatively young and therefore more readily able to adapt to climate change adversity. Approximately 52% of the population are aged 0–24 and 37% of the population are between the ages of 25–54. However, there are multiple climate change induced threats that will affect health, human development and food security. These threats include the following.

1. The average temperature in Egypt, as a result of global warming, is predicted to increase by 4 °C in Cairo and by 3.1 to 4.7 °C in the rest of Egypt by 2060.[9] The IPPC has projected a 2–11 °F (1.4 to 5.8 °C) rise in average global surface temperature during the twenty-first century.
2. Annual precipitation may drop by 10 to 40% over most of Egypt by 2100.
3. The Mediterranean Sea level is predicted to rise by 0.5 m by 2050. "This will lead to flooding the coastal areas along the Nile Delta."[10]

Extreme water scarcity will be one of the most visible and difficult consequences of climate change for the Egyptian population. Moreover, the current water stress will be further exacerbated by significantly rising household and irrigation demands as the population surges from 94 million in 2017 to 130 million in 2030. "Nile River provides more than 95% of all water to Egypt. The annual rainfall varies from a maximum of 180 mm/year on the North coast, to an average of 20 mm on the middle of Egypt to 2 mm/year on the Upper Egypt. Both water supply and demand are expected to be exaggerated by climate change. It is expected by 2050 that climate change will raise water demand by an average of 5%."[11] Water scarcity will in turn affect crop production and food security under climatic conditions that are already exacerbating heat levels and drought conditions. "A number of countries in North Africa already face semi-arid conditions that make agriculture challenging, and climate change will be likely to reduce the length of growing season as well as force large regions of marginal agriculture out of production. Projected reductions in yield in some countries could be as much as 50% by 2020, and crop net revenues could fall by as much as 90% by 2100, with small-scale farmers."[12]

A study by the Egyptian Society for Migration Studies calculated that "the decline in agriculture activities due to temperature increases is expected to range from 10% to 60%. The production of the strategic crops will achieve significant reduction by the middle of the century (2050) as the following: Production of wheat will reduce by 18%; Production of rice will reduce by 11%; Production of maize will reduce by 19%. Egypt is among the high potential countries/regions for food crisis during the coming 40 years."[13] As a developing country with a large and significantly growing population, Egypt will be called upon to respond aggressively with adaption and mitigation programs to forestall the dramatic effects of sea level rise, coastal flooding, drought episodes, health challenges and food insecurity that will be certainties in the twenty-first century.

Climate Change Refugees

The main factor inducing climate refugee flows will be sea level rise and flooding. A secondary threat is drought and cultivation challenges that force farmers, their families and farm migrants to relocate. As noted in a World Bank report, the Nile Delta is subsiding at a rate of 3–5 mm per year thus a rise of 1.0 m would flood one-quarter of the Nile Delta, forcing about 10.5% of Egypt's population from their homes. A report examining sea level rise in developing countries had dire predictions for Egypt. The "Egyptian coastal population are undeniably exposed to the effects of SLR, with its accompanying flooding as the population is expected to double before the year 2050, if the present growth rate is maintained. SLR is expected to affect Egypt in many ways; with just a one-meter rise in the Mediterranean Sea, the Nile delta stands to suffer tremendously; 6.1 million people are predicted to be displaced and 4500 square kilometers of cropland will be lost."[14]

As one of the more developed states in the region, Egypt is unlikely to experience notable external climate refugee pressures, however internal migration will likely be pronounced particularly to the metropolis of Cairo. Yet Cairo and indeed all of Egypt will face the burden of extreme water scarcity, which will have a deciding impact upon migration patterns. "Egypt is facing an annual water deficit of around 7 billion cubic meters. By the year 2020, Egypt will be consuming 20% more water than it has and the United Nations is already warning that Egypt could run out of water by the year 2025."[15] As water security affects all levels of society, with profound influences upon health and agriculture production and sustainability, the burden of Egypt's deepening water crisis will be an immense challenge for the government and population.

City Profile: Alexandria

Alexandria, located on the north coast, is the second largest city in Egypt with a population of 5 million. It is a historic city in antiquity and often referred to as the "The Pearl of the Mediterranean." The main climate change risk to the city is sea level rise and flooding with the attendant problem of saltwater intrusion. The IPCC predicts that the Mediterranean will rise 30 cm to 1 m in the twenty-first century. An Arab Forum for Environment and Development report stated that "forty percent of Egyptian industry is located in Alexandria alone; a 0.25 meter rise in sea level would put 60 per cent of Alexandria's population of 4 million below sea level, as well as 56.1 per cent of Alexandria's industrial sector. A rise of 0.5 m would be even more disastrous, placing 67% of the population, 65.9% of the industrial sector, and 75.9% of the service sector below sea level. Thirty percent of the city's area would be destroyed, 1.5 million people would have to be evacuated, and over 195,000 jobs would be lost."[16]

Aside from the tragic human costs associated with dramatic sea level rise, the economic burdens on Egypt from sea level rise and flooding will be prohibitive and push back years of socio-economic development. "Based on the 0.5-m scenario, estimated losses of land, installations, and tourism will exceed US$32.5 billion." An average business loss is estimated at USD 127 million per year because most tourist facilities such as hotels, camps and youth hotels are located within 200–300 m of the shoreline. It has been widely reported that 8 million people would be displaced in Egypt by a 1 m rise in sea level, assuming no protection and existing population levels.[17]

Innovation and Adaption Strategies

1. In 2016 Egypt signed the Paris Agreement.
2. In 2009, Law 9 is promulgated. It is an amendment and update to the 1994 Environmental Law 4 which addresses the protection of the environment and a national action plan.
3. In 2009 the Climate Change Risk Management Programme is launched.
4. In 2008 the Ministry of Interior enacted Traffic Law no. 121 of 2008, which directs that all passenger transport vehicles (taxis, buses and microbuses) more than 20 years old will not be permitted to renew their operating licenses, effective August 1, 2008.
5. In 2005 Egypt ratified the Kyoto Protocol, which was signed in 1999.
6. In 2005 the Egyptian Designated National Authority for the Clean Development Mechanism is established.
7. In 1999 Egypt's Climate Change Action Plan is established.
8. In 1999 the Initial National Communication on Climate Change is initiated.
9. In 1997 the inter-ministerial National Climate Change Committee is launched by the Egyptian government.
10. In 1994 the UNFCC is ratified.

Israel

Israel is a small, semi-arid country with a population of 8.1 million that borders the Mediterranean Sea, Egypt and Lebanon. The land area is 20,770 km^2 and the coastline is 440 km^2. In 2016, the GDP was USD 311 billion, HDI rank 18 and the country contributed approximately 0.2% to global emissions. The main climate change threats include drought, heat waves, heat related illness, sand storms, air pollution, sea level rise, groundwater pollution and diminishing water security. Increased heat, increased drought conditions, desertification and declining rainfall are projected with certainty for the country in the twenty-first century. Although water scarcity is a certainty facing Israel, advances in desalinization technology—for which the country is a recognized leader—may help to mitigate aspects of the crisis. There is evidence to suggest a that 50% reduction in rainfall in the Jordan River basin will occur. The relatively young population of Israel will support adaption and mitigation efforts and perhaps minimize some of the health implications of climate change. In 2017, approximately 28% of the population were aged 0–14; 15.6% were aged 15–24; and 37.2% were aged 25–54.

Flooding and Drought

Although flooding is a less significant problem it is likely to increase due to the effects of sea level rise. A report by the government of Israel noted with concern that a "trend of rising seawater levels, totaling more than 10 cm, was recorded in the Mediterranean Sea over the past two decades, consistent with scenarios which range from 1 to 10 cm per decade. Such a rise is associated with increased flooding along the coastal plain and increased intrusion of seawater to the coastal aquifer which leads to salinization. Wave storms with wave heights exceeding 3.5 meters have also increased along with exceptional storms with a wave height of more than 6 meters, which are expected to pose major risks to coastal installations and to the collapse of the coastal cliff."[18]

The inter-related crises of drought, declining rainfall and water insecurity are troubling signs on the horizon for Israel. Israel experienced several years of continuous drought between 2004 and 2016. A 2016 NASA study reported that the worst drought in 900 years, which started in 1998, affected the eastern Mediterranean and countries of Cyprus, Israel, Jordan, Lebanon, Palestine, Syria and Turkey. Israel has a rainfall range of a mere 20–30 inches (50–70 cm) annually. Climate forecasts from the government indicate "an increase in average annual temperatures of 0.3–0.5 °C per decade, a reduction in rainfall of

1.1–3.7%, an increase in the frequency and intensity of heat waves and extreme events such as floods and droughts, and an increase in the probability of forest fires in the Mediterranean region over the next fifty years."[19]

The effect of drought conditions that become exacerbated will have significant consequences on a range of health and socio-economic conditions. Although the country has suffered water shortages since its inception in 1948, and regionally for centuries, the problem is becoming more pronounced. "By the mid-1990s, after several multi-year cycles of drought and over-pumping of natural water reservoirs, the situation developed into a crisis so severe that an adequate supply of domestic water was in jeopardy. The current cumulative deficit in Israel's renewable water resources, according to the Hydrologic Service, amounts to approximately 1 billion cubic meters (1000 MCM), which is half of country's annual consumption."[20] Accordingly, as 70% of fresh water sources are traditionally utilized for agriculture irrigation, the depletion of water reserves will increasingly impact food security.

Extreme Weather

Extreme weather is a climate phenomenon that Israel will be forced to address with varying degrees of mitigation and adaption success. Extreme heat that is forecasted over the twenty-first century will aggravate drought, diminish scarce water sources and profoundly effect health. The National Oceanic and Atmospheric Administration (NOAA) defines a heat wave as: "a period of abnormally and uncomfortably hot and unusually humid weather, typically, a heat wave lasts two or more days."[21]

Israel's 2010 Climate Change Report also noted the extreme weather events facing the country and the destructive consequences upon the nations' fragile water sources. "The Ministry of Environmental Protection refers to several impacts including: increase in the frequency and severity of floods, which may cause major damage to property and people, twenty five percent reduction in water availability for 2070–2099 in comparison to 1961–1990; reduction in groundwater recharge; loss of an estimated 16 MCM of water for each kilometer along the coastal plain as a result of a potential rise in sea level of 50 centimeters and changes in the salinity level of the Sea of Galilee."[22]

Health, Human Development and Food Security

In addition to fresh water scarcity, the main challenges posed by climate change will be elevated temperatures, heat waves, drought, sea level rise and related health complications. Heat stress in the population of Israel will become pronounced. "Heat-related illnesses (e.g., heat cramps, heat exhaustion, heat syncope, heatstroke or heat rashes) can occur when high ambient temperatures overcome the body's natural ability to dissipate heat. Older adults, young children and persons with chronic medical conditions are particularly susceptible to these illnesses and are at high risk for heat-related mortality."[23] The child and youth population (aged 0–24 years) is 52.9% and the elderly (65 and older) population is approximately 8.1%. Older citizens often experience chronic respiratory problems and have lower lung and breathing capacity. Conversely, as the immune system is not generally fully formed until the age of 19, the high proportion of children and young people in Israel are a concern for the government and health officials.

A study on the effects of heat on the number of emergency room (ER) visits in Israel observed that the contribution of mean daily temperature to the number of ER visits was small but significant and the study further noted that the number of visits to ERs increased by 1.47% per 1 °C increment in ambient temperature.[24] As heat levels predictably rise, the health of thousands of citizens will be placed at risk. There is a clear correlation between air pollution and respiratory problems, which can be exacerbated by extreme heat. For example, asthma attacks and other respiratory diseases are expected to increase during heat waves due to interactions with air pollution in general and from a rise in the frequency of wildfires.[25,26]

A government report on the effects of climate change enumerated a range of concerns confronting the population. "Climate change is expected to affect public health directly through physical influences such as extreme heat and cold events and indirectly through the effect on chronic and infectious diseases and on mortality and morbidity from external sources. Extreme weather conditions are known to increase the frequency of certain illnesses, such as cardiovascular and respiratory diseases, while climate change is known to affect the presence and behavior of vector-borne diseases. Other climate factors, such as ultra-violet radiation, are associated with such diseases as cancer and cataracts."[27]

Approximately 50% of Israeli farmland is threatened by climate change. Water is essential for irrigation, thus in the face of rising temperatures and drought, Israeli officials remain challenged to secure water for personal, agricultural and business use. Successive governments over many decades have demonstrated admirable conservation methods and in recent years have shown a sound commitment to water enhancing technologies including desalinization. In general, "farming accounts for 70% of global water usage, to feed 9 billion people by 2050, we will need 60% greater agricultural production and a 15% increase in water withdrawals, Farming livestock is up to 10 times more intensive than farming crops—protein-focused diets are increasingly impacting water supplies"[28]

An examination of the climate induced threats facing Israel point strongly to a range of issues that must be addressed with proactive adaption and mitigation strategies. The Israel National Report on Climate Change stated the following agriculture and food security issues that confront the nation:

- damage to crops due to a reduction in water availability in the soil;
- a 20% increase in water demand for irrigation;
- a reduction in fruit and vegetable yields;
- the emergence of new pests and pathogens and an increase in the frequency of animal and plant diseases;
- sharp cutbacks in allocations of freshwater resources for agricultural irrigation;
- an increased risk of soil erosion;
- reduced productivity of farm animals;
- shortage of animal feed and an increase in its cost;

- shortening of the productivity season of pastureland;
- and damage to populations of pollinating insect species.[29]

Climate Change Refugees

As a relatively prosperous developed country, Israel will be in a stronger position than most to mitigate the adverse effects of climate change. Although external climate refugee migration is not expected, the likelihood of internal migration to avoid severe climate conditions is high. The primary threats to the population are drought and extreme water scarcity. These problems pose significant challenges for the government and policy planners. One mitigating and hopeful sign is that Israel is a leader in recycled water systems and desalinization technology.

Innovation and Adaption Strategies

1. In 2016 Israel signed the Paris Agreement.
2. In 2016 the Israeli government approved a plan to reduce GHG emissions and raise energy efficiency. The government has targeted a reduction per capita in GHG emissions to 7.7 tons of CO_2 equivalent by 2030, which will translate into a 26% change from emissions in 2005. Israel has submitted the plan to the UNFCCC.
3. In 2015 the Israeli cabinet approved Government Decision No. 542 that set a national per capita GHG emissions reduction target of 8.8 tons of CO_2 equivalent by 2025 and 7.7 tons of CO_2 equivalent by 2030. Included in the decision are the following targets: A 17% reduction in electricity consumption by 2030; and 17% of the electricity generated in 2030 will be from renewable sources. In 2017, approximately 3% of Israel's electricity was generated by renewables.
4. In 2011 the Ministry of Environmental Protection established the Israel Climate Change Information Center.
5. In 2009 the government approved Decision No. 250, to appoint an inter-ministerial committee to address issues related to environmental protection and climate change adaption.

6. In 2009 the government approved Decision No. 474, which established a strategic plan for climate change adaption.
7. In 2007 a National Water Desalinization Plan was launched with a goal to expand natural water sources by 35%.
8. In 2006 the Ministry of Environmental Protection established an inter-ministerial steering committee with the goal of advancing recommendations on climate change adaption.
9. In 2002 the National Master Plan was created to address the severe water shortage crisis facing the country.
10. The UNFCCC was ratified in 1996.

Jordan

Jordan is a semi-arid country with a population of 7.7 million (2017) and generously accommodates an additional 2 million Syrian refugees. The land area is 88,802 km^2 with a sea coast of 540 km. It shares borders with Iraq, Israel, Saudi Arabia, Syria and the West Bank and is situated along the Red Sea. The main climate change threat is drought, with a minor threat emanating from sea level rise and flooding on its small border with the Red Sea. The country also experiences periodic earthquakes, which scientists attest can in some instances be triggered by climate change. The GDP is USD 39 billion (2016) and Jordan's HDI ranking is 80. The Jordanian population is relatively young, which is a positive mitigating factor against some of the health problems linked to climate change. Approximately 13% of the population is under 4 years of age, 25% is under 14 years of age, 60% of the population is aged 15–64 years and 3.5% is over 65 years old.

Flooding and Drought

Although drought is the main climate change threat facing the country, there is a flood risk due to sea level rise affecting the Red Sea. Jordan's small coastline is still vulnerable to anticipated flooding in the twenty-first century that will cause devastation to the coastal population. According to a recent report, "the vulnerability assessment showed clearly that the geographically restricted coastline of the Gulf of Aqaba is vulnerable to climate change impact. The high vulnerability was noticed for main exposure determinants which are: sea level rise, increase of sea surface temperature and CO_2 concentrations. This is true as the Gulf of Aqaba contains sensitive ecosystems and habitats, which are very vulnerable to any changes in sea composition."[30]

Historically, drought has been a significant problem for Jordan and according to climate data projections will be exacerbated over the coming century. Approximately 80% of the country is uninhabited desert. In recent years, the 2007–2009 drought affected nearly 5 million people and caused damage to over half of the country's cropland. Even without the impact of climate change drying trends, Jordan is an arid country with severe water scarcity problems, thus projected warming temperatures and prolonged droughts will be severe challenges for the population and government. "As one of the four driest countries in the world, water scarcity is by far the greatest impediment to planned growth and development, with far-reaching impacts across all sectors. Water scarcity is expected to be exacerbated by climate change, which has already decreased rainfall levels and increased temperatures. In response to concerns about water resources, the Jordanian government proposes to increase the use of treated wastewater to supplement irrigation rather than rely on potable water."[31]

Projections for the country related to drought and diminished water security capacity include the following: "Rise in annual maximum temperature of up to 5.1 °C and rise in annual minimum temperature of 3.8 °C by 2085 (warming is stronger during the summer), some models project temperatures to rise evenly across the country while others suggest the increase will be strongest in the eastern and southern regions; increase in the frequency of heat waves, 10-day increase in the number of consecutive dry days from 2040–2070 (increase will be greatest in the southern Aqaba region), precipitation projections are highly variable but point to an overall decrease between 15–60 percent from 2011 to 2099."[32]

Extreme Weather

Extreme weather events such as prolonged elevated temperatures, heat waves, sandstorms, extended drought and erratic rainfall patterns seriously affect the country. Data from the Second National Communication to UNFCCC and IPCC–AR5 conclude with climate projections that appear certain. These include:

1. a predicted rise in mean temperature for all of the country;
2. extremely likely warmer summers compared to other seasons;
3. extremely likely heat waves where the analysis of summer temperature, monthly values and the inter-annual variability reveal that some thresholds could be exceeded, especially in summer months where the average of maximum temperature for the whole country could exceed 42–44 °C;
4. in 2100, the country will be subjected to a temperature increase of 3–5 °C. Representative comprehensive pathways (RCp) 4.5, predicts that a rise of up to 2.1 °C (1.7–3.2 °C) in the projected maximum temperature is extremely likely;
5. more heat waves. An analysis of summer temperature monthly values and the inter-annual variability reveals that some thresholds could be exceeded. A pessimistic but possible projection for the summer months predicts that the average of maximum temperatures for the whole country could exceed 42–44 °C.[33]

Jordan is also situated in an active earthquake zone, which creates further vulnerabilities as climate change can induce volcanic and earthquake activity. Noted researcher Bill McGuire has concluded from groundbreaking research studies that, "anthropogenic climate change does not simply involve the atmosphere and hydrosphere, but can also elicit a response from the Earth's crust and mantle. In this regard, we hope that it will encourage further research into those mechanisms by which climate change may drive potentially hazardous geological and geomorphological activity, and into the future ramifications for society and the economy."[34]

Health, Human Development and Food Security

Water scarcity and drought are reinforcing phenomena that have profound consequences upon society and all measures of development. As one of the countries in the world with the most severe water scarcity scenario, Jordan must forge ahead with aggressive mitigation and adaption strategies. The 2016–2025 National Water Strategy is an essential commitment to address the critical situation for citizens in terms of both planning and programs. As noted in the National Water Strategy report, "Jordan's renewable water resources are limited and insufficient to meet national demand. There are growing signs of apparent overuse in an increasing number of watersheds and aquifers. Jordan's annual renewable resources of less than 100 m^3/Capita are far below the global threshold of severe water scarcity of 500 m^3/Capita. National water resources and water balance are facing negative impacts due to higher demand, over abstraction and the effects of climate change."[35] Water levels per capita were 3600 m^3 per year in 1946 and 145 m^3 per year by 2008 due to population growth and climate change and will be reduced to 90 m^3 per year by 2020 if no substantive action is taken.[36]

A government analysis observed multiple health threats to the population related to climate change impacts. According to the report:

- Climate change will increase micro-organism growth; leading to increases in water and food-borne diseases; in contrast flooding—which is the result of extreme rainfall through concentrating the annual rainfall in a small interval—leads to the disruption of water purification and contamination with sewage disposal systems, leading to an increase in the probability of epidemics due to vector-, water- and food-borne diseases.
- Climate change may also influence the seasonal pattern of respiratory diseases, cardiovascular diseases and mortality.
- "In general the main climate change exposure issues for the health sector in Jordan are drought, dust or sand storms, decreasing precipitation, rising temperatures, flooding due to extreme rainfall and shifting in the rainy season."[37]

Moreover, analyses based upon eight climate data models have also projected severe drought and water scarcity scenarios. "For three periods (2020–2050, 2040–2070, and 2070–2100) there is significant increase in temperature and hence in evaporation. Also, the data suggest reduced precipitation and hence, drought. In conclusion, Jordan's water sector is extremely vulnerable to climate change, especially to temperature increase, decrease in precipitation and increase of evapotranspiration in the area of study."[38]

By 2017, Jordan resettled the second highest of number of refugees per 1000 inhabitants in the world and was the sixth highest refugee receiving nation. The country has a current refugee population of 2.9 million, primarily from Syria followed by Palestine, which is equivalent to 30% of the population. For decades, the humanitarian support from and sacrifice of the Jordanian people and government for the refugee community in the country has been extraordinary. One of the well observed implications of climate change is the disproportionate burden placed upon the poor, marginalized, infirm and elderly. With approximately 80% of the country covered by desert, only 10% of land suitable for crop cultivation; and water tables projected to decline to a crisis level of 90 m^3 per year by 2020, Jordan will, perhaps more than most countries, experience profound climate change challenges that adversely impact health, human development and economic opportunity.

Climate Change Refugees

The realities of severe prolonged drought and water scarcity will increasingly endanger health, agriculture and economic livelihoods in the country and will, in all eventuality, pressure thousands of Jordanians to migrate internally and in some cases externally. Political factors and the on-going tragedy in Syria will also be significant factors in the migration scenario.

Innovation and Adaption Strategies

1. 2016 Paris Agreement signed;
2. 2016–2025 National Water Strategy. Core elements of the strategy include: the Water Sector Capital Investment Program (2016–2025);

the Water Demand Management Policy; Energy Efficiency and Renewable Energy in the water sector policy; Water Substitution and Re-Use Policy; the Water Reallocation Policy; the Surface Water Utilization Policy; the Groundwater Sustainability Policy; the Climate Change Policy for a Resilient Water Sector; the Decentralized Wastewater Management Policy; and the Action Plan to Reduce Water Sector Losses;
3. 2015 Intended Nationally Determined Contribution;
4. 2014 Third National Communication on Climate Change;
5. 2013 Climate Change Adaptation and Low Emission Development Strategy;
6. 2009 Second National Communication on Climate Change;
7. 2008–2022 Water for Life: Jordan's Water Strategy;
8. 2006 National Strategy and Action Plan to Combat Desertification;
9. 1999 First National Communication on Climate Change;
10. 1996 Government National Environment Action Plan (NEAP).

Interview

Dr. Khaled Hassan is a demographer and economic expert with a track record in the analysis of climate changes in the Middle East and North Africa region. He holds a Ph.D. in economics management. He is the vice-president of the Egyptian Society for Migration Studies. He has published numerous books, studies and academic articles.

Can you describe your work related to climate change?

My work concentrates on scientific research, studies and analyses of negative phenomena of climate change. I look at the impact of climate change on the population and the population's economic and social characteristics and activities, and what limits its ability to achieve its goals in sustainable development.

What concerns you the most about climate change in your country?

I presented a scientific research paper entitled "The Future Impacts of Climate Change on Egyptian Population" at the 27th IUSSP International Population Conference, Busan, South Korea, August 26–31, 2013.

In this study I point to sea level rise and flooding of a large part of the Nile Delta as the greatest negative impact of climate change in Egypt.

The Nile Delta is currently subsiding at a rate of 3–5 mm per year. A rise of 1.0 m would flood one-quarter of the Nile Delta (this may occur within the next 25 to 33 years according to the current rate of subsiding), forcing about 10.5% of Egypt's population from their homes (World Bank, 2009, and UNDP, 2013). This impact would be even greater combined with the expectation of a doubling of Egypt's population to about 160 million by 2050—the current population density in the Nile Delta is about 4000 people per square mile.

As a result of sea level rise and flooding of a large part of the Nile Delta (the most cultivated part of Egypt's land), food production and supply would be effected. Nearly half of Egypt's crops, including wheat, bananas and rice, are grown in the delta. On the other hand, the remaining areas of the Nile Delta (not under water) would be affected by saltwater from the Mediterranean Sea, which would contaminate the fresh groundwater used for irrigation. Agricultural activities and self-sufficiency of food would be also exposed to a temperature increase. The decline in agricultural activities due to a temperature increase is expected to be 10–60% (UNDP, 2013). Egypt is one of the countries highly likely to experience a food crisis in the next 40 years.

Sea level rise will affect the population's economic activities. A rise of 0.5 m would be even more disastrous, placing 67% of the Alexandria's population of 4 million, 65.9% of the industrial sector, and 75.9% of the service sector below sea level. Some 30% of the area of the city of Alexandria would be destroyed, 1.5 million people would have to be evacuated and more than 195,000 jobs would be lost. Alexandria is not the only Egyptian city that would be devastated by a 0.5 m rise in sea level. Other cities threatened by a rising sea level in the delta include Port Said, Matruh and Arish.

Sea level rise will harm Egypt's tourism sector. The Nile Delta is home to much of Egypt's tourism. For cities like Alexandria and Matruh, the threat of rising sea level will reduce their capability to sustain tourism as well as the desire of tourists to visit them. Some 49% of Alexandria's tourism industry would be under water if sea level rose by 0.5 m.

What concerns you the most about climate change globally?

From my point of view, the most concerning thing about climate change is global warming. As we all know, it is a global phenomenon that manifests in warming of the Earth's atmosphere. This is due to the excessive rise of GHGs (CO_2, water vapor, methane, ozone gas and chlorofluorocarbons), which maybe a throwback to the Industrial Revolution and the resulting gases.

Global warming has caused a hole in the ozone layer. Chlorofluorocarbons are the biggest contributor to the ozone holes. Consequently, the Earth receives harmful rays from the Sun through the hole (such as UV rays). GHGs contribute to acid rain. Acid rain is harmful to plants, humans and the entire ecosystem. Human beings have contributed to the creation of GHGs by burning fossil fuels, deforestation and other environmentally harmful activities.

Tracking and analyzing global efforts to reduce global warming is more of a concern to me than tracking the causes and consequences of global warming. Efforts to reduce global warming can be classified into international and individual areas. The most appropriate of these efforts to be applied and expanded on are as follows:

1. Reduce the use of fossil fuels and expand the use of clean energy from its various resources (solar energy, wind, and water slopes).
2. Increase vegetation, conserve forests and plant more trees—they have a role in the elimination of CO_2 and mitigation and cooling of the atmosphere.
3. As individuals we can conserve energy, use public transport rather than increase the number of private cars on the road and build well-insulated homes so that we do not need winter heating and summer cooling.
4. The best solution is still "the cooperation and the Union of the States of the world in the face of global warming that threatens our lives and stays on the ground."

Can you describe a specific sustainability project in your area or country that is contributing to a "greener world" and lower carbon footprint?

Egypt is implementing a number of projects aimed at adopting green economy policies. These projects were designed to suit Egypt's environmental and economic priorities, and to achieve sustainable development while implementing the National Strategy for Sustainable Development 2030.

Bio-Energy Project

The bio-energy project is a sustainable rural development project funded by the Ministry of Environment in collaboration with the Global Environment Facility and the UNDP, in coordination with some relevant ministries, including the ministries of petroleum, electricity and energy, international cooperation and agriculture, as well as the Social Fund for Development and the New Energy Development Authority. The bio-energy project aims to the use bio-energy technologies, create a new market for them in Egypt and encourage young people to enter the market as entrepreneurs to provide this service in the governorates of Egypt.

The project has established 1000 bio-gas units and seeks to establish 250 thousand units in the long term, while achieving a number of objectives including recycling animal waste and reducing the burning of agricultural waste and the use of fossil fuels. The bio-gas produced by these units is used in cooking and lighting. The project implemented the largest commercial units of bio-gas production in Egypt, which produces 50 m^3 of bio-gas each day in a farm in Fayoum governorate.

Wind Power Project

Egypt has adopted a long-term plan for wind energy, aiming to provide 20% of its electricity needs from renewable energy sources by 2020, 12% of them from wind energy. In 2010, Egypt received USD 1.3 billion to invest in clean energy development through projects related to solar, thermal and wind energy.

Waste Recycling Project

The waste recycling project is an important model, recycling municipal and solid waste and converting it into energy, bio-fuel or solid fuel for use in cement and/or ceramics factories. In this regard, the Ministry of Environment has examined 22 proposed sites for the establishment and classification of these projects in terms of size and operational capacity as pilot projects managed with the participation of the private sector.

These projects may change the map of energy production and management of wastes in Egypt for the next 30 years. The field of waste recycling can provide 15 job opportunities to turn over 1 ton of solid waste and 25 job opportunities to produce 1 ton of bio-fuel.

Transport Sustainability Project

This project aims to reduce carbon emissions and global warming, increase energy saving and solve traffic congestion problems in the streets of Egypt. The cost of the project is USD 43 million, of which USD 7 million comes from the UNDP and the remainder is funded by the Egyptian government and private sector partnerships. The project provides public transport buses, which are characterized by the comfort and luxury they offer to passengers. New cities will be linked with Cairo to encourage mass transport.

What major new steps could your government take to mitigate the growing threats from climate change?

New, major governmental steps to mitigate the growing threats from climate change can be summarized as follows:

1. expanding the process of constructing sea bumpers to reduce the impact of sea level rise and to protect the beaches along the Mediterranean Sea;
2. more effective implementation of policies to reduce population growth and reach the replacement level of 2.1 by 2020;
3. increasing the populated area of Egypt, to reduce the high population density in the delta and basin of the Nile and reduce the pressure on services and facilities, through the establishment of new cities with comprehensive infrastructure and facilities to meet the current and future demands for housing;
4. Oil and natural gas are the main sources of energy in Egypt to date, serving 95% of Egypt's total energy needs. As a result of its energy deficit, Egypt will remain an importer of oil and gas over the next few years, according to Egypt's Energy Strategy 2030 and its on-going modernization reaching to the year 2035. Such a situation represents an additional challenge to the Egyptian economy, as it becomes vulnerable to price disturbances in global energy markets and the depletion of Egypt's foreign exchange resources affecting the trade balance and reducing the competitiveness of the national economy. Therefore, diversification of energy sources is being reconsidered to maximize the use of domestic resources, which are characterized by sustainability and price stability. As a result of these challenges, successive governments of Egypt since 2013 have directed great attention to electricity

production from renewable sources (solar energy, wind and water), taking into account Egypt's wealth in these resources. Egypt's Energy Strategy 2030 aim is to generate 20% of total electricity production in Egypt from renewable energy resources, including hydropower, by 2020, increasing to 25% by 2030. Of course, these targets have a positive impact on the environment and are consistent with the plans of the state to reduce emissions and maximize the use of domestic resources from renewable sources of energy, in addition to the production of electricity from nuclear power from the station at Dabaa in 2024. The target capacity for the production of electricity from renewable energy is about 4300 MW, distributed between 2300 MW of solar energy and 2000 MW of wind energy. This is encouraged by the steady decline in the cost of electricity production from solar energy, which has dropped from USD 359 per megawatt-hour in 2009 to USD 79 in 2014, or 78% in five years. It is also one of the cleanest sources of energy. Production of electricity from solar energy enhances self-sufficiency and alleviates dependence on oil-exporting countries. It also limits and significantly reduces the cost of transporting fuel to generating stations and then transferring electricity generated from production areas to consumption areas. The sunlight falling on a house is sufficient to meet a family's energy needs. Add to this many other important uses of solar energy, such as desalination of sea water, heating homes, lighting and cooking.

Can you describe a particular area of climate change that you are concentrating upon with research or a grassroots development project.

I am currently studying the inter-relation and interaction between global climate change and the achievement of the sustainable development goals 2015–2030. On September 25, 2015, the efforts by UN bodies crystallized in the development of 17 goals represented in their entirety by the sustainable development agenda from 2015 and until 2030. Sustainable Development Goals (SDGs) are a global framework to coordinate efforts around ending poverty and hunger, combating inequality and disease, ensuring education for all, gender equality and female empowerment, protecting the planet and ensuring prosperity for all. Each goal has specific targets to be achieved over the next 15 years (UN, 2015).

Some of these objectives have been addressed during the period of the Millennium Development Goals 2000–2015, where the lives of millions have been saved and the living conditions of others have been improved and developed, according to the data and analysis of the UN report on the Millennium Development Goals in 2015.

The inter-relations and interactions of many of the sustainable development goals are notable. For example, the issues of poverty will not be solved in isolation from achieving successes in the areas of the spread of education and health services, equality and empowerment of women and other goals. Combating climate change and its impacts has captured the world's attention in the last three decades and is capped as one the global SDGs of 2015–2030.

There is no doubt that climate change in general and global warming in particular have had a negative effect, with varying and uncertain degrees, on the ability of countries to achieve the SDGs. Current global warming rates may exacerbate economic and social challenges, especially in developing countries that depend on resources that are sensitive to climate change.

The study will address the current and future effects of climate change and global warming on the efforts of countries, especially the developing countries in achieving some of the SDGs. More specifically, the study will see the effects of climate change and global warming on hunger and extreme poverty, universal basic education, gender equality, female empowerment, health and child mortality.

What is your reaction to the UNIPCC.s accepted calculation by Prof. Myers of a minimum of 200 million climate change refugees by 2050 and the world's readiness to cope with this humanitarian and environmental challenge.

I am convinced by the idea of growing numbers of refugees and displaced persons due to climate change and expected environmental disasters, but I disagree with the numbers adopted by the IPCC, based on Prof. Myers' calculations for a number of reasons.

We should distinguish between three basic definitions in migration studies, which are: who is the migrant, who is the refugee and who is the displaced person. A migrant is a person who has moved voluntarily from their country of origin to another country (international migration) or who has moved voluntarily from one place to other place inside the same country (internal migration), this type of migration is usually done by individuals for personal reasons that are mostly economic or social. A refugee is a person who has moved from one country to another for involuntary reasons (such a person is called a forced migrant or international refugee). The reasons for forced movement may be political (due to wars and/or civil and national disputes), cultural or religious (due to religious, ethnic and cultural persecution) or environmental (due to environmental and climatic change disasters). This type of movement usually involves a mass exodus. A displaced person moves involuntarily from one place to another inside the same country. They are called internally displaced persons. The reasons and types of displacement are the same as for refugees.

The current estimation of the number of international migrants is about 250 million, which represents about 3.6% of the total world population (currently estimated at 7.349 billion). The estimated number of refugees due to environmental and climate change crises is about 27 million (which represents about 10.8% of all international migrants).

According to the Norwegian Refugee Council, the number of people displaced by armed conflict and natural disasters in the world reached 31.1 million in 2016. According to the report, 24 million people were affected by natural disasters last year, and the majority were forced to leave their homes. In China, 7.4 million people have been displaced by natural disasters, 5.9 million in the Philippines, 2.4 million in India and 1.2 million in Indonesia.

Accordingly, the estimate to the number of refugees and displaced persons is about 58.5 million of the 7.349 billion as a world population. Estimates indicate that world population will reach about 9.0 billion people by 2050. Consequently, the estimated number of refugees and displaced persons due to environmental and climate change reasons are subject to one of the following three scenarios:

1. The optimistic scenario: As a result of growing environmental awareness among individuals and governments, the commitment of countries to the content of environmental conservation and climate change agreements and with the continuation of global population growth

and its concentration in poor countries where citizens are more vulnerable to the adverse impacts of climate change, the expected number of asylum and displaced persons due to climate change risks will reach 71 million by 2050.
2. The moderate scenario: Assuming stabilization of environmental degradation and climatic disasters at the current level, with an increase in global population and its concentration in developing and poor countries where citizens are more vulnerable to the negative effects and risks of climate change, the number of people exposed to asylum and displacement is expected to increase by 15–20%, to reach about 82–85 million by 2050.
3. The pessimistic scenario: Assuming a continuous deterioration in environmental conditions, the absence of international policies or programs to reduce the impact of environmental degradation and the negative effects of climate change, a lack of commitment, especially in developed countries, to international obligations in this regard, and a growing global population concentrated in poor countries where citizens are more vulnerable to the adverse effects of climate change, the number of asylum and displaced persons is expected to increase by 25–30%, reaching 89–92 million by 2050.

I am convinced that there will be growing numbers of refugees and displaced people due to climate change and environmental disasters. This analysis shows the large gap between my estimates and those of the IPCC, based on Prof. Myers' calculations.

Interview

Dr. Mehmood Ul-Hassan leads Capacity Development at the World Agroforestry Center to address development problems. He holds a Ph.D. from the University of Bonn, Germany, and studied trans-disciplinarity in research organizations. He has 25 years' experience in international agricultural research in Asia, Africa and Europe, and has more than 30 refereed publications to his credit.

Can you describe your work related to climate change?

I head the Capacity Development Unit at the World Agroforestry Center. The unit works towards developing individual and system capacities in developing nations for sustainable agricultural development, where trees and agroforests play a critical role in farming landscapes and people's livelihoods, food and nutritional security.

What concerns you the most about climate change in your country?

Well, we work in more than 30 developing countries across Asia, Africa and Latin America. The most common issue among all these countries is their national readiness to deliver on their climate and sustainable development commitments. Many developing countries do not understand the implications of their commitments and what it will take in terms of knowledge, skills and attitude changes required to successfully deliver on their commitments. Besides, the policymaking is rarely evidence based or risk informed, and hence the interventions proposed generate more problems than they solve. There is lack of enabling environment (weak or lacking appropriate institutions, implementation arrangements, assessment tools, financing) and institutional capacities have serious gaps.

What concerns you the most about climate change impacts upon agriculture?

Most climate related research in many developing countries is based on only four major commodities (rice, wheat, maize and potatoes), which tends to cause monoculturalization of agriculture, which has been one of the underlying causes of agricultural contribution to emissions. The world needs to shift towards perennialization of agricultural production systems (having crops that have a sustainable yield for several years, minimizing agricultural operations). Instead of improving crop productivity through solutions that regulate microclimates on farms (e.g., through maintaining tree cover on and around farms), much of the research emphasis is on varietal improvements for temperature, drought and disease tolerance. This situation is exactly what has led to increased agricultural contributions to emissions.

Can you describe a specific sustainability project in your area or country that is contributing to a "greener world" and lower carbon footprint?

There are several projects of interest: Stakeholder Approach towards Risk Informed Decision Making; The ASB Partnership for the Tropical

Forest Margins (www.asb.cgiar.org); and Propoor Rewards for Environmental Services in Africa (www.presa.worldagroforestry.org). More information on other projects can be found on the following websites:

- Rewarding Upland Poor for Environmental Services (www.rupes.worldagroforestry.org);
- http://www.worldagroforestry.org/output/taking-heat-out-farming-0;
- http://www.worldagroforestry.org/output/restoring-lives-and-landscapes;
- http://www.worldagroforestry.org/.

What steps could your government take to mitigate the growing threats from climate change?

The governments of developing countries need to invest in reforming and capacitating their research, implementation and policymaking structures to identify and address issues in reality for their own sake (not to satisfy donors). Developed countries need to reform their institutions to provide financing and expertise in developing such capacities without political considerations. Today's climate financing windows are highly politicized and do not optimally generate the results that we urgently need.

What is your reaction to the UNIPCC's accepted calculation by Prof. Myers of a minimum of 200 million climate change refugees by 2050 and the world's readiness to cope with this humanitarian and environmental challenge?

This is a conservative estimate in my opinion. I foresee between 500 million to 1 billion climate refugees by 2050! The world isn't ready to tackle even fewer than 100 million due to current wars in the Middle East.

Can you describe a particular area of climate change that you are concentrating upon with research or a grassroots development project?

I am looking at how small-scale tree/agroforest growers can participate in climate finance initiatives. This work is taking place in Cameroon.

Notes

1. "Looming Crisis of the Much-Decreased Fresh-Water Supply to Egypt's Nile Delta" 13 March 2017 GSA Release No. 17–11. The Geological Association of America (GSA).
2. Arab Forum for Environment and Development, 2010.
3. EJF (2009) "No Place Like Home—Where next for climate refugees?" Environmental Justice Foundation: London.
4. Neumann B., Vafeidis A.T., Zimmermann J., Nicholls R.J. (2015) Future Coastal Population Growth and Exposure to Sea-Level Rise and Coastal Flooding—A Global Assessment. PLoS ONE 10(3): e0118571. pmid:25760037.
5. Climate: Observations, projections and impacts: Egypt, Met Office.
6. Leila Radhouane, "Climate change impacts on North African countries and on some Tunisian economic sectors" Journal of Agriculture and Environment for International Development—JAEID 2013: 101–113.
7. Egyptian Meteorological Authority, 17–20 Oct 2014, "Drought condition and management strategies in Egypt, Tamer A. Nada.
8. Vizy, E.K. and K.H. Cook, 2012: Mid-twenty-first-century changes in extreme events over northern and tropical Africa. Journal of Climate, 25(17), 5748–5767.
9. Taeleb, 1999.
10. The Future Impacts of Climate Change on Egyptian Population by Khaled El-Sayed Hassan, Economic Demographer and Statistical Expert, Egyptian Society for Migration Studies, 2013.
11. Eid, 1999.
12. Boko, M., I. Niang, A. Nyong, C. Vogel, A. Githeko, M. Medany, B. Osman-Elasha, R. Tabo and P. Yanda, 2012: IPCC, 2012: Managing the Risks of Extreme Events and Disasters to Advance Climate Change Adaptation. A Special Report of Working Groups I and II of the Intergovernmental Panel on Climate Change [Field, C.B., V. Barros, T.F. Stocker, D. Qin, D.J. Dokken, K.L. Ebi, M.D. Mastrandrea, K.J. Mach, G.-K. Plattner, S.K. Allen, M. Tignor, and P.M. Midgley (eds.)]. Cambridge University Press, Cambridge, UK, and New York, NY, USA, 582 pp.
13. The Future Impacts of Climate Change on Egyptian Population by Khaled El-Sayed Hassan, Economic Demographer and Statistical Expert, Egyptian Society for Migration Studies, 2013.

14. Dasgupta, S., B. Laplante, C. Meisner, D. Wheeler, and J. Yan (2007) The Impact of Sea Level Rise on Developing Countries; A Comparative Analysis. World Bank Policy Research Working Paper 4136, Washington: Development Research Group, World Bank. 2007.
15. "Egypt's Water Crisis—Recipe for Disaster" Amir Dakkak, EcoMena, January 4, 2016.
16. Khaled El-Sayed Hassan, Egyptian Society for Migration Studies and Arab Forum for Environment and Development (AFED) 2010.
17. Broadus et al., 1986; Milliman et al., 1989; IPCC, 1997—R.T. Watson, M.C. Zinyowera, R.H. Moss (Eds) Cambridge University Press, UK.
18. "Adaptation to climate change in Israel Recommendations and knowledge gaps" State of Israel, Ministry of Environmental Protection, Office of the Chief Scientist, February, 2014.
19. Ibid.
20. CSD-16/17 Drought and Arid Land Water Management, CSD-16/17 National Report Israel 1/11, Drought and Arid Land Water Management, Government Focal Points: Michael Zaide, Strategic Planning Engineer, Planning Division, Water Authority, Ministry of National Infrastructure, Israel.
21. NOAA. National Weather Global Service- Glossary. http://w1.weather.gov/glossary/index.php?letter=h.
22. "Climate Change Report" Ministry of Environmental Protection, 2010 and Israel Water Authority.
23. MMWR: Heat-Related Deaths—United States. 1999, http://www.cdc.gov/mmwr/preview/mmwrhtml/mm5529a2.htm, 2003.
24. Novikov I., Kalter-Leibovici O., Chetrit A., Stav N., Epstein Y. Weather conditions and visits to the medical wing of emergency rooms in a metropolitan area during the warm season in Israel: a predictive model. Int J Biometeorol. 2012;56:121–127. https://doi.org/10.1007/s00484-011-0403-z.
25. Vynne C., Doppelt B. Climate Change Health Preparedness in Oregon: An Assessment of Awareness. Environmental Health Committee: Preparation and Resource Needs for Potential Public Health Risks Associated with Climate Change. Climate Leadership Initiative Institute for a Sustainable Environment University of Oregon & the Oregon Coalition of Local Health Officials; 2009.
26. Schwartz B.S. Climate change and public health. Medscape Public Health; 2008. http://www.medscape.org/viewarticle/574087.

27. "Adaptation to climate change in Israel Recommendations and knowledge gaps" State of Israel, Ministry of Environmental Protection, Office of the Chief Scientist, February, 2014.
28. Tamar Weiss, "Can Israeli Technology Help Beat the Food-Security Threat Posed by the Global Water Crisis?" SNC, December 28, 2016.
29. Israel National Report on Climate Change, Ministry of Environmental Protection Pinhas Alpert, Prof., Department of Geophysics and Planetary Sciences, Unit of Atmospheric Sciences, Tel Aviv, 2011.
30. Jordan's Third national Communication on Climate Change, Submitted to The United Nations Framework Convention on Climate Change (UNFCCC) 2014.
31. USAID Fact Sheet, Climate Change Risk Profile Jordan, March 2017.
32. Ibid.
33. Jordan's Third National Communication on Climate Change, Submitted to The United Nations Framework Convention on Climate Change (UNFCCC). 2014.
34. Bill McGuire, "Climate forcing of geological and geomorphological hazards", Phil. Trans. R. Soc. A 2010 368 2311–2315; https://doi.org/10.1098/rsta.2010.0077. Published 19 April 2010.
35. National Water Strategy of Jordan, 2016–2025, Ministry of Water and Irrigation, Hashemite Kingdom of Jordan, 2016.
36. USAID Fact Sheet, Climate Change Risk Profile Jordan March 2017.
37. Jordan's Third National Communication on Climate Change, Submitted to The United Nations Framework Convention on Climate Change (UNFCCC) 2014.
38. Ibid.

9

Europe: UK, Italy, Greece

"What need does the Earth have of us?"
Pope Francis, Laudato Si': On the Care of Our Common Home

Introduction

The European continent faces weather extremes from Arctic conditions in the north to drought episodes in the south. Flood episodes and heat waves have been noted with increasing regularity in recent decades. Invariably European countries face a multitude of climate change impacts that are diverse and vary across countries. According to an environmental report there are significant climate change repercussions facing Europe. The more serious consequences include: "Increased risk of inland flash floods, more frequent coastal flooding and increased erosion. Mountainous regions will suffer from glacier retreat, reduced snow cover and declining winter tourism, and extensive species losses (in some areas up to 60% under high emissions scenarios by 2080). Southern Europe: high temperatures and drought are projected to worsen in a region already vulnerable to climate variability. Reduced water availability, hydropower potential, summer tourism and, in general, crop productivity are predicted. Increase in health risks due to heat waves and the frequency of wildfires."[1] The IPCC in a 2014 landmark report warned of multiple climate change adversities that will strike Europe. These include, "retreating glaciers, sea level rise, longer growing seasons, species range shifts, and heat wave-related health impacts. Future impacts of climate change will likely negatively affect nearly all European regions, with adverse social, health, and infrastructure effects. Many economic sectors, such as agriculture and energy, could face challenges."[2]

UK

The UK is a leading global economy and island nation with an extensive coastline, which imperils the population to the dramatic consequences of sea level rise, saltwater intrusion and flooding. The country is situated between the North Atlantic Ocean and the North Sea and has a land area of 243,610 km^2, a coastline of 12,429 km and a population of 65,511,098 (2017). The GDP is USD 2.629 trillion (2016) and HDI rank is 16 (2016). In 2016, CO_2 emissions were 381 million tons, which represents a 5.8% decline from the previous year. The demographic profile of the UK is favorable to effective climate change mitigation and adaption strategies based upon two compelling factors: a high standard of living and a relatively young population. The age demographic (2016) includes 0–14 years: 17.44%; 15–24 years: 12.15%; 25–54 years: 40.74%; and 65 years and older: 17.9%. In general, the main climate change threats facing the country are flooding from sea level rise and elevated levels of rainfall in the twenty-first century.

Flooding and Drought

The proximity of the UK to the Arctic will expose the population to increased levels of flooding due to the rapid rate of sea ice melt and general global warming. As noted, studies on the acceleration of Arctic sea ice melt indicate an escalating trend and warming rates faster than previously anticipated. Some studies suggest up to 90% of Arctic sea ice will disappear by 2070. A recent British government report noted with concern that, "The impacts of flooding and coastal change in the UK are already significant and expected to increase as a result of climate change. Warming of 4 °C or more implies inevitable increases in flood risk across all UK regions even in the most ambitious adaptation scenarios considered. Climate change is projected to reduce the amount of water in the environment that can be sustainably withdrawn whilst increasing the demand for irrigation during the driest months."[3]

The predicted human dislocation, health crises and infrastructure damage from flooding will be significant and intensify as the twenty-first century progresses. The impact from flooding in coastal zones has far reaching implications that affect human life, disease levels, infrastructure, economic livelihoods employment and the allocation of government resources, which will be will further stretched as climate change mitigation and adaption costs escalate. The Environment Agency (EA) estimates that 5.2 million properties in the UK (one in six properties) are at risk of flooding. More than 5 million people live and work in the 2.4 million properties that are at risk of flooding from rivers or the sea; 1 million of these properties are also at risk of surface water flooding and a further 2.8 million properties are susceptible to surface water flooding.[4]

A collaborative report by the World Health Organization, UNFCCC and the British government cited the main climate change predictions that will emerge with a significant degree of certainty in the event that emissions are not appreciably curbed. These findings include:

1. under a high emissions scenario, mean annual temperature is projected to rise by about 3.7 °C on average from 1990 to 2100;
2. under a high emissions scenario, an average of 585,400 people are projected to be affected by flooding due to sea level rise every year between 2070 and 2100;
3. it is likely that vectors (ticks and mosquitoes) will spread within the UK due to warmer summers, wetter springs and milder winters by the 2080s;

4. under a high emissions scenario, the number of days with very heavy precipitation (20 mm or more) could increase by just over 10 days on average from 1990 to 2100, increasing the risk of floods;
5. under a high emissions scenario, the number of days of warm spells is projected to increase from about 10 days in 1990 to just over 165 days on average in 2100.[5]

Extreme Weather

The UK has a long history of windstorms, heavy rainfall and flood episodes. As climate change occurrences escalate, these scenarios are expected to increase in intensity. The floods in June–July 2007 caused electrical damage and power disruption to 50,000 people and left up to a million citizens without drinking water and resulted in an estimated GBP 2 billion in damages. In 2015, extreme rainfall affected northwestern England, Ireland, Scotland and Wales. On December 4, 2015, an Atlantic storm deposited 13.44 inches of rainfall upon the inhabitants Honister Pass in Cumbria, which set a national record for rainfall in one 24 hour period.

The UK Met Office reported that the winter of 2015/2016 was the second wettest since records began in 1910, having "record-rainfall totals" accompanied by "exceptional warmth."[6] "In a recent climate modelling study researchers found that climate change made the UK's record December rainfall in 2015, which caused the devastating floods, 50–75% more likely."[7] In contrast, heat levels in the UK are expected to increase. A government climate change report observed that heat waves in the UK, like the one experienced in 2003, are expected to become the norm in summer by the 2040s.[8] This scenario will add to the risk of additional flooding, mudslides and spreading mosquito and other insect problems.

Health, Human Rights and Food Security

As noted the main climate change threats facing the UK are coastal flooding and high rainfall patterns that will increase exponentially over the twenty-first century. This in turn will generate a host of challenges for human health, socio-economic development, infrastructure security and government budgets. An extensive report examining health consequences from climate variations noted the following eventualities that are likely to impact citizens.

1. Heat-related mortality is projected to increase steeply in the UK in the twenty-first century. We estimate this increase to be approximately 70% in the 2020s, 260% in the 2050s, and 540% in the 2080s, compared with the 2000s heat-related mortality baseline of around 2000 premature deaths, in the absence of any physiological or behavioral adaptation of the population to higher temperatures.
2. Cold-related mortality is projected to remain substantially higher than heat-related mortality in the first half of the twenty-first century. However, it is estimated to decline by 2% in the 2050s and by 12% in the 2080s, compared with the 2000s baseline. The South East, London, East and West Midlands, the East of England and the South West appear to be most vulnerable to current and future effects of hot weather. The elderly, particularly those over 85 years of age, are much more vulnerable to extreme heat and cold compared with younger age groups. Future health burdens may be amplified by an aging population in the UK.[9]

Food security is a prevalent issue in the context of climate change and is expected to remain relatively stable for the UK population. However, due to rising heat and drought conditions in the southern European zone and the impact upon growers in the affected EU countries, higher food prices are a certainty. Moreover, water insecurity in southern Europe will also have an appreciable effect upon irrigation and cultivation that will have a subsidiary impact upon food production, security and pricing. The burden will be disproportionately high upon citizens living in poverty including those on fixed incomes.

Perhaps the most significant effect of climate adversity will be in the form of coastal flooding. As an island region, the UK will be at the forefront of global flood impacted populations. An assessment of sea level rise noted the devastating outcomes that will confront the UK. The assessment showed that 10–15% of the UK's coastline is comprised of 10 km long stretches that are below 5 m elevation and that 3009 km (16%) is subject to erosion. The study also calculated that 69% of GDP is located within 50 km of the coast and that 78% of the country's population live within this zone. At present, 414,000 people are exposed to sea level rise in the UK.[10] Moreover, scientific projections indicate an acceleration of sea level rise conditions for the foreseeable future. "The frequency of extreme events around the UK, resulting from increases in time-mean sea-level rise, can be expected to increase by a factor of more than 10 at many locations, and at some locations by more than 100 over the next century. This applies both to the moderate extremes at lower return periods, such as the annual maximum water level, and to the often much more damaging 1-in-100-year events."[11] Accordingly the government and affected coastal populations will have to be prepared for considerable adaption and mitigation responses including wide-scale migration inland to higher flood protected locations.

Climate Change Refugees

The UK is a prosperous economic union situated in the top five global economic ranking. Moreover, a relatively young population will serve to mitigate some of the challenges of climate change dislocation. The country is not expected to generate climate change refugee flows, however substantial coastal flooding will increase exponentially over the course of the twenty-first century, which will force thousands of citizens to relocate inland.

Innovation and Adaption Strategies

1. 2017 Policy Paper, UK Climate Change Risk Assessment.
2. 2016 Paris Agreement signed.
3. The Climate Change (Scotland) Act 2009 sets forth the goal for Scotland to meet a 42% reduction in emissions by 2020 and yearly reductions between 2010 and 2050.
4. The Low Carbon Transition Plan (2009) sets forth the procedures for attaining carbon reductions as required under the Climate Change Act. The target is to reduce emissions by more than 33% in order to

meet the 2020 reduction target in the act—to reduce emissions by at least 34% below the 1990 baseline.
5. The Climate Change Act (2008) is the government strategy to reduce GHG emissions. The act aims to reduce GHG emissions by at least 80% of 1990 levels by 2050.
6. The Renewable Energy Strategy (2009) sets forth the strategy to increase the use of renewable energy for heat, electricity and transport to meet the target to ensure 15% of energy is derived from renewable sources by 2020.
7. 2008 Climate Investment Funds, established by the World Bank, provide funding to support projects and targets in developing countries. Funds are contributed by 14 donor nations: the UK, Australia, Canada, Denmark, France, Germany, Japan, Korea, the Netherlands, Norway, Spain, Switzerland, Sweden and the USA. In 2016, USD 8.3 billion in pledged donations were reserved for projects such as renewable energy, energy efficiency, sustainable transportation and climate resilience programs in 72 pilot nations.

Italy

Italy is hot, semi-tropical country and a top ten global economy with a GDP standing of USD 1.852 trillion (2016), a population of 63 million (2017), and an HDI rank of 26 (2016). The climate is Mediterranean with Alpine features in the north and hot, dry conditions in the south, which will intensify over the coming century. The total land area is 301,340 km^2 with a coastline of 7468 km. Of the 21 zones in the country 15 are coastal. Climate change and environmental challenges include coastal flooding, saltwater intrusion, land slides, mudflows, avalanches, earthquakes, volcanic activity and land subsidence in Venice, a city that is confronted with elevated flood crises that will increase exponentially in the twenty-first century. The country has a relatively young population that will contribute to effective climate change adaption. In 2016 the age demographic was: 0–14 years: 13.69%; 15–24 years: 9.74%; and 25–54 years: 42.46%. The senior population of 65 years and over is 21.37%. The country share of global CO_2 emissions in 2016 was 0.9%.

Flooding and Drought

Coastal flooding and drought conditions in southern Italy with attendant heat waves will be the dual challenges facing the country in the coming decades. These problems will escalate in intensity with each successive decade. Recent scientific projections conclude that extreme sea level events will increase (high confidence) mainly dominated by the global mean sea level increase.[12] "The most vulnerable areas are the coasts of Toscana, the Tevere mouth (Lazio), the southern part of Lazio, the Volturno estuary and the coast south of Salerno (particularly in Cilento) in Campania and Sicily. The Northern Adriatic basin is particularly at risk due to the presence of the Po delta (Emilia Romagna) and the Venice lagoon (Veneto). In this area, the coastline is rarely more than 2 meters wide and due to subsidence various zones presently lie below sea level."[13] The Italian Ministry for the Environment predicts that the flooding crisis facing the country will be significant. The Ministry report observed that for the Italian coasts, sea level rise will imply high risks. About 4500 km^2 of coastal areas and plains would be at risk of coastal flooding in 2080 (according to a study carried out by NASA-Goddard Institute for Space Studies); floods might occur in northern Italy (upper Adriatic Sea), central Italy (the coastline between Ancona and Pescara, the coasts near Rome and Naples) and in southern Italy (Gulf of Manfredonia, coastline between Taranto and Brindisi, eastern and southern Sicily).[14] A technical report from the Ministry for the Environment, Land and Sea quantifies the areas with a high risk of flooding: they cover an area of 7.774 km^2, corresponding to 2.6% of the national territory. Floods can cause several consequences to human health, infrastructures and the environment. "The most dramatic floods in Italy occurred in the Po (1951, 1994, and 2000) and Arno river basin (1966)."[15]

It is interesting to note the significant level of climate change alteration that will affect mountain ranges and the interconnected rivers, streams and agricultural fields. In Europe, a vivid example is the European Alps which spans eight countries, Austria, France, Germany, Italy, Liechtenstein, Monaco, Slovenia and Switzerland, and have a combined population of 14 million people who live in mountainous communities. Over the past century, the Alps have been affected by a very high temperature increase of almost 2 °C, more than twice the rate of average warming recorded in the northern hemisphere. A wetting trend in the north Alpine region and a drying trend in the south have been observed.[16]

A European Environment Agency (EEA) report examining impacts upon the entire ecology and human activity of the Alps region due to climate change noted extensive impacts that will increase exponentially in the twenty-first century. Moreover, similar to the Tibetan Plateau, climate scientists have observed a warming and glacier melt ratio that is higher than other regions of the world. This observed and quantified environmental reality is the subject of intense examination by climate scientists. According to the EEA report, "In the coming decades, global change will alter the hydrological cycle of the Alps, probably leading to more droughts in summer, especially in the southern Alps, and more floods and landslides in winter, especially in the northern Alps. This will also affect the amount of water that the water towers of Europe can provide to million[s of] people in lowland areas and the economic sectors relying on water such [as] agriculture, hydropower production, industry, winter tourism, river navigation. Global warming could accelerate the reduction in snow cover at low altitudes as well as the melting of glaciers and permafrost at higher altitudes thus increasing the hydro-geological risks in the mountain areas."[17]

As noted, Italy will also face consequential warming trends over the next century that will create extended drought conditions with multiple socio-economic and health effects. A scientific team led by Giovanni Forzieri and Luc Feyen calculated the drought risk to the European continent by utilizing scenario analysis and integrated modeling tools. The conclusions are a blunt reminder of the severe consequences facing the continent from one aspect of climate change. The researchers concluded that the Iberian Peninsula, Italy and the Balkan region will face the most severe drought impacts. They further reported that "by using a large ensemble of climate projections originating from various combinations of 12 global and regional climate models, we capture a wide range of future climate developments and their variability and that drought conditions will intensify as the twenty-first century progresses, with up to a 40 percent reduction in minimum stream flow by the 2080s."[18] Drought conditions will further jeopardize water security and related urgencies such as irrigation. The water security crisis will pose substantial health and economic burdens upon Italy. Accordingly, research by the European Commission has already signaled the dimensions of the crisis. "By the end of the century, southern Europe—including the Iberian Peninsula and Italy—will have 80 per cent more droughts than at present. The European Commission predicts that water shortages will be made worse by population growth and increased demand. It said that in the past 30 years, dry spells have cost Europe over €100 billion (or £83 billion)."[19]

Extreme Weather

Extreme weather is a recurring environmental phenomenon in Italy and is expected to increase in severity. In 2015, intense summer heat waves, a tornado in Venice, and floods across the country caused substantial hardship and economic damage. Moreover, ten of the warmest years in Italy since 1800 occurred after 1990. Climate scientists project that this trend will continue, if not escalate. This will have a devastating and residual effect upon human health, agriculture and water security. Vulnerable groups such as the elderly and health impaired will have particular difficulties. Doctor visits and hospital admissions will rise dramatically. "During the last 30 years, the Italian mean temperature anomaly (a measure of how the average temperature in Italy in any given year differs from a historical, multi-year average) was almost always higher than the global over land air temperature anomaly. In 2013, the mean temperature anomaly was +1.04 °C in Italy, compared to the global mean of +0.88 °C."[20] Increased flood episodes will correspondingly contribute to more landslide incidents. Thus the number of citizens who are endangered and ultimately evacuated due to landslides is expected to rise. "The population potentially exposed to landslides every year is 995,484 people spread out over 21,182 km^2 (7% of the national territory). In 2013, 112 major landslide events were recorded. It is estimated that every year 6,153, 860 people are exposed to the effects of flooding in Italy. Floods in Italy killed 1557 people from 1951 to 2013."[21]

Health, Human Rights and Food Security

A recent IPCC report summarized the multiple climate change threats in southern Europe that affect many sectors:

- Climate change is expected to impede economic activity in southern Europe more than in other sub-regions (medium confidence).
- Southern Europe is particularly vulnerable to climate change (high confidence), as multiple sectors will be adversely affected (tourism, agriculture, forestry, infrastructure, energy, population health) (high confidence).
- Heat-related deaths and injuries are likely to increase, particularly in southern Europe (medium confidence).[22]

Italy is a major agricultural producer. Major production includes: cereals, vegetables, horticultural products, potatoes, fruit, olive oil and wine. Cultivation of all of these agriculture products is weather dependent and vulnerable to the increased heat and water scarcity implications of climate change in the country. A 2014 analysis of environmental impacts upon Italian agriculture noted a number of challenging scenarios. The report observed that, "Italy and southern France show a reduction of rain in summer time. The combined effect of significant increase in temperature and reduction in rainfalls has determined an increasing irrigation demand and has contribute[d] to [a] rise [in] water deficit.[23] Water shortage represents the most important consequence of these meteorological phenomena on EU Southern countries agricultural production. For these reasons increased plant heat stress was recorded in Spain, Italy and in the Black Sea area like Turkey. In these countries agricultural sector absolutely must improve its water use efficiency to counter the costs associated with the increased use of this input."[24]

Climate Change Refugees

Italy is not expected to generate external climate refugee flows. The relative wealth of the country, leading global economic position and prominence within the G20 alliance of nations will mitigate some of the challenging costs associated with climate change prevention and adaption strategies. However, the country will not be immune to some of the harsh realities of climate dislocation of populations and infrastructure damage. The main environmental threats facing the Italian population are drought conditions, primarily in the south, extensive sea level rise, coastal flooding, severe infrastructure damage and related health challenges. With extensive Mediterranean and Adriatic coastlines, the anticipated flooding that will increase in severity over the twenty-first century will force hundreds of thousands of citizens to migrate inland.

City Profile: Venice

The historic city of Venice, situated on the Adriatic, is severely affected by sea level rise. Periodic flooding has been a constant in Venetian history. The city rests on a series of islands in a region that is forecasted to experience significantly enhanced sea level rise and flooding in the twenty-first century. In 1900, the city's main square was submerged repeatedly and in 1996 there were 99 floods. In September 2016, four retractable gates were placed into the Malamocco inlet, a seaway into the Venice Lagoon that is designed to protect 57 flood barriers that can be elevated at high tide. The completion date for all the gates is June 2018. The urgency of flood mitigation and adaption is clear to Venetians and the municipal government. In 2013, a sea level rise of 1.43 m caused flooding and water damage to 50% of the city. The crisis facing the city was observed in a recent policy report. "Venice is one of the most vulnerable parts of the country with regard to flooding and extreme weather events. The city is located along the Venice lagoon, the largest wetland of the Mediterranean Sea. The total surface area of the lagoon is about 550 km^2 of which 420 km^2 is directly subject to the sea tides. In this area, high tides have increased in terms of frequency and intensity due to the low elevation, a rise in sea level as well as a reduction in land level (subsidence). As a result, the area of Venice is losing land to the sea."[25]

Innovation and Adaption Strategies

1. 2017 National Adaptation Plan.
2. 2016 Paris Agreement signed.
3. 2014 National Communication on UNFCCC submitted.
4. Italy has a reduction of GHG emissions target of 13% by 2020 based on 2005 emissions. The goal is codified in Law n. 96/2010, June 4, 2010.
5. A target set in the EU 2020 Climate and Energy Package, adopted in 2009, aims for national legal requirements including a 20% reduction in greenhouse gas emissions in the EU by 2020 compared to 1990 levels.
6. A 2030 Commitment: a 40% domestic reduction in emissions below 1990 levels by 2030, to be fulfilled jointly by EU member states; 13% reduction in domestic emissions below 2005 levels by 2020, in accordance with Decision No. 406/2009/EC of the European Parliament and Council.
7. 2003 Project MOSE initiated to erect a storm surge barrier that could close the lagoon of Venice during high flood periods in the Adriatic Sea.
8. A fund for GHG emissions reduction and energy efficiency (Finance Law 2001 Article 10) (2001) is set up to address the reduction of atmospheric emissions and to encourage energy efficiency and sustainable energy sources.
9. 1994 UNFCCC: signature and ratification (April 15, 1994).

Greece

Greece is a historic country with an ancient civilization. The country has a population of 11.1 million (2017), a land area of 131,957 km^2 (50,949 square miles) with a coastline of 13,676 km and borders the Aegean, Ionian and Mediterranean seas. The climate is dry and hot with a topography dominated by mountains and fertile valleys that support productive agriculture mainly centered upon barley, corn, sugar beets, wheat, beef and fruit. The country has more than 1400 islands, which are increasingly threatened by sea level rise and flooding as are the extensive coastal areas of Greece. The GDP per capita is USD 19,100 and the country is ranked 29 in the global HDI positioning of 188 nations. CO_2 emissions in Greece have dropped substantially from 7.86 metric tons in 1996 to 6.23 metric tons in 2015. The serious climate change threats facing the country include drought, desertification, sea level rise and coastal flooding. Indeed, the burdens facing the population and government will be substantial as adaption and mitigation action become increasingly urgent. A comprehensive Bank of Greece report detailing the climate change costs facing the country warned that "on a general note, the impact of climate change on all sectors of the economy that were examined was found to be negative and, in several cases, extremely so."[26]

Flooding and Drought

For Greece, the primary climate change threats include rising temperatures, heat waves, drought, desertification in some regions and sea level rise that will affect coastal communities with moderate to severe intensity. The southern Mediterranean region in general will face substantial threats in the twenty-first century, and in most scenarios, such as drought, escalating with each successive decade. With regard to sea level rise and flooding, a study by the Greek Ministry of the Environment observed that, "Mean sea level in the Mediterranean is expected to rise at the rate of 5 cm/decade. In particular sea level will rise about 50 cm by the year 2100 (with an uncertainty range of 20–86 cm). Delta Nile, Venice and Thessalonica appear to be the more sensitive areas in the Mediterranean."[27] The effects of sea level rise and flooding are dramatic and extend far beyond the immediate crisis of flooding. Additional burdens include health challenges, economic losses, disruption to economic livelihoods, infrastructure damage, diverted public expenditures and forced migration. "According to a projection based on a 0.5 m sea level rise by 2100, 15% of the current total area of coastal wetlands in Greece is expected to be flooded. The estimated economic losses from erosion (for land uses: urban, tourist, wetland, forest and agricultural) for the entire Greek territory for 2100 amounts approximately €356 million and €649 million for 0.5 m and 1 m sea level rise, respectively."[28]

Unfortunately for Greece, many of the vulnerabilities are due to topography and existing geographic formations that cannot be changed. The country has the 11th longest coastline in the world at 13,676 km (8498 miles) thus the combined phenomenon of Arctic ice melting, sea level rise, coastal flooding and saltwater intrusion pose inevitable challenges. "About 20% of Greece's total coastline is ranked as being of moderate-to-high vulnerability to developments likely to arise on the basis of projections. Both the long-term change in sea level and extreme, transitory events will affect many sectors of the economy, including tourism, land use and transportation. The total cost of anthropogenic contribution to sea level rise will come to tens of millions of euros each year."[29] In addition to coastal flooding there is also the perennial threat of torrential rains and flash floods. On October 26, 2014, for example, Greek authorities were compelled to confront serious reconstruction challenges after heavy rainfall led to flash floods in Athens that caused extensive damage to homes, infrastructure and trees.

The other major threat facing the nation is drought, which will intensify with each successive decade in the twenty-first century and likely beyond. The southern Mediterranean region is projected to undergo significant warming patterns and declining rainfall. A government report noted with a sense of urgency the crises that confront the country. "The greatest increases in drought periods are projected for the eastern part of the mainland and for Northern Crete, where 20 more drought days are expected per year in 2021–2050 and up to 40 more drought days are expected in 2071–2100. The number of days with a very high risk of fire is expected to increase significantly by 40 in 2071–2100 across Eastern Greece (from Thrace down to the Peloponnese), while smaller increases are expected in Western Greece. The impact of climate change on all sectors of the economy that were examined was found to be negative and, in several cases, extremely so."[30]

Drought carries forth many symptoms dangerous both to human health and economic sustainability. Health related heat stroke, respiratory problems, water scarcity, irrigation challenges, desertification, crop reduction and forest fires are just some of the serious challenges confronting the country. Indeed, in August 2007, forest fires were at their worst levels in many decades and illustrated the country's vulnerability to drought and heat induced fires. According to recent projections, there was an "increase in the number of days with highly increased forest fire risk in eastern Greece: 20 days in 2021–2050 and 40 days in 2071–2100."[31]

According to a general consensus of estimates, over the next century the country is facing: an increased temperature of 2.5 °C compared to the measurement period of 1961–1990 period, a 20 to 50 cm rise in sea level and a rainfall decline of 12%, which will seriously impact agriculture and water security. A 2017 report noted the severity of the heat/drought crisis facing southern European nations such as Greece. "All European regions are vulnerable to climate change, but some regions will experience more negative impacts than others. Southern and south-eastern Europe is projected to be a climate change hotspot, as it is expected to face the highest number of adverse impacts. This region is already experiencing large increases in heat extremes and decreases in precipitation and river flows, which have heightened the risk of more severe droughts, lower crop yields, biodiversity loss and forest fires."[32]

The recent financial struggles of the Greek government and people will constrain their ability to respond aggressively to climate change adaption and mitigation demands that will escalate in the coming decades. However, as the data and case studies illustrate from numerous government and international reports, waiting and delay is no longer an option for the people of Greece. Successive Greek governments will increasingly be called upon to respond with bold and pragmatic solutions to address climate change threats that will only intensify. The great Greek civilization, the wellspring of tremendous innovation through history, will be called upon to meet the test of environmental leadership in the twenty-first century in what may well prove to be its most serious national crisis.

Extreme Weather

Extreme weather is a phenomenon that affects all countries to varying degrees and is commonly identified as one of the symptoms of climate change. In Greece, extreme weather patterns are primarily related to elevated heat and rainstorms. These environmental events, particularly heat, will escalate over the century. In July 2017, for example, several regions in central and north central Greece experienced temperatures of more than 40 °C. The heat wave was the worst since 2007 when an intense heat episode struck the country. In contrast, in June 2017, rainstorms affected several Greek regions, and hailstorms episodes damaged crops and farmland.

A report analyzing anticipated extreme weather impacts for Greece noted the following scenarios. "It is estimated that, even under the intermediate scenario A1B, the Greek mainland in 2071–2100 would, compared to now, have some 35–40 more days with a maximum daily temperature of 35 °C or more, while even greater would be the increase by around 50 at the national level) in the number of tropical nights (when minimum temperatures do not fall below 20 °C". Changes are also expected in precipitation extremes. "In Eastern Central Greece and NW Macedonia, the maximum amount of precipitation occurring within 3-day periods is expected to increase by as much as 30%, whereas in Western Greece it is expected to decrease by as much as 20%."[33]

In July 2015, Greece was again beset by untenable heat. The temperatures exceeded 36 °C nearly every day. Moreover, March 2015 was the third warmest March in recent history after 2010 and 2002. There is a pattern whereby typical high heat periods, such as July and August are expanding into other months. The IPCC has even hypothesized about the prospect of "October summers" in Greece. A recent scientific report examining heat wave projections in Greece calculated a scenario of escalating heat extremes. "Results showed widespread significant changes in temperature extremes associated with projected warming in the near (2031–2050) and the far (2071–2100) future under SRES [special report on emissions scenarios] A1B, especially for those indices derived from daily minimum temperature. The findings of this study give evidence that extreme events such as summer 2007 will be more common and frequent in the future."[34]

Health, Human Rights and Food Security

For Greek citizens, the immediate health impact of rising heat levels and heat waves will be heat stroke, air pollution, respiratory problems, complications for people with coronary disease, rising levels of skin cancer and interrupted water security. Moreover, as with any health challenge, the burdens are disproportionately experienced by people in vulnerable situations such as the elderly, children, the impoverished and disabled. Approximately 15% of the population are 0–14 years old and 20% are 65 years old and above. A stagnant birthrate and rapidly aging population pose significant health policy challenges for Greek society.

Moreover, there are major challenges that need to be addressed with respect to climate change impacts. As studies indicate, "for populations in the EU, mortality has been estimated to increase 1–4% for each one-degree increase of temperature above a cut-off point. During heatwaves, deaths increase from a range of causes. Elderly people are most at risk because ageing impairs the body's physiological capacity to regulate its own temperature (thermoregulation). Children, people with chronic diseases, and those confined to bed, need particular care during extremely hot weather."[35] The Projection of Economic Impacts of Climate Change in Sectors of the European Union Based on Bottom-up Analysis project estimates 86,000 extra deaths per year in EU countries with a global mean temperature increase of 3 °C in 2071–2100 relative to 1961–1990.[36]

The combined effects of climate change will have a disruptive and costly impact upon the agriculture sector and society in general. In one study it was reported that, "climate change is predicted to have significant negative impacts on many fields of activities, for example on biodiversity loss, species and habitat range shifts, agriculture, forestry, fisheries, tourism, transport, activities in coastal areas and the built environment in urban centers, mainly due to increased temperature, drought, extreme weather events and rising sea levels. The cost of reducing emissions and adapting to climate change for the Greek economy is estimated to be around 500 billion euros by 2100, which, if no measures are taken, may rise to almost 700 billion euros."[37] Projections suggest devastating reductions in agricultural lands due to climate change induced heat, drought, desertification and sea level rise. "One study calculated a loss of agricultural land of 19% by 2040–2050 and 38% by 2090–2100; negative impacts in southern Greece and islands including Crete, and negative effects that will be most pronounced near the end of the twenty-first century."[38]

Water is essential for human life and development. In climate change scenarios where water security is threatened by extreme heat and drought, the results for a society can be devastating. Greece, according to the World Resources Institute, is projected to have severe water stress by 2040. Escalating temperatures and symptoms such as drought and desertification will accelerate the water scarcity crisis in Greece. Recent studies indicate a considerable drop in precipitation levels for the nation over the twenty-first century. "Water resources of the eastern Mediterranean and Middle East region have been investigated using a high-resolution regional climate model (PRECIS) by comparing precipitation simulations of 2040–2069 and 2070–2099 with 1961–1990 (29). Greece is expected to have an 18% precipitation decrease by midcentury, and 22% by the end of the century."[39]

Climate Change Refugees

Greece will experience significant climate change impacts related to extreme heat, drought episodes that escalate, desertification, flooding and degrees of coastal flooding depending upon the region. The country is not expected to generate climate refugee flows. However, citizens affected by the more severe impacts of climate change, such as chronic extreme heat, desertification and localized coastal flooding, will be compelled to relocate. A more serious scenario will be migration pressure from other countries, notably countries in Africa, where climate change impacts range from moderate to severe. Traditionally, Greece is seen as a gateway for refugees and migrants fleeing hardship to enter a highly developed economy and also gain potential entry to other EU countries. The global climate refugee movement will escalate with intensity in the twenty-first century. Estimates range from 200 million to 500 million by 2100. Predictably, the migration pressure to enter developed countries will be intense. Thus countries such as Greece, which are relatively close to highly vulnerable climate impacted populations, will inevitably experience intense climate refugee migration pressures.

Innovation and Adaption Strategies

1. In 2016, signed the Paris Agreement.
2. In 2016 the first National Strategy for Adapting to Climate Change was initiated in Law 4414/2016.
3. Law 3851/2010 provides for renewable energy incentives. It primarily addresses wind power, photovoltaic systems, hydroelectric power, solar power, geothermal energy and bio-mass areas.
4. Law 3889/2010 provides funding opportunities for environmental initiatives and support for the Green Fund, which receives support from tax revenues, the EU and other donation sources.
5. Law 3831/2010 regulates taxes on different categories of vehicles to encourage low emissions. Hybrid and low-emission vehicles are exempt.
6. Directive 2009/29/EC provides compensation to carbon intensive industries exposed to carbon leakage caused by the indirect costs of the EU Emissions Trading policy.
7. Law 3661/2008 promotes initiatives to introduce energy efficiency into buildings. Under the law, energy performance certificates are compulsory for all new buildings. The law also provides for studies and licensing procedures for buildings to include a study on passive solar system heating/cooling/electricity production systems that utilize renewable energy.
8. In 2005 the Central Water Agency was established. Its goal is to develop a national water policy for Greece.
9. Law 3423/2005 promotes the introduction of bio-fuels and other renewable fuels into the Greek economy.

Interview

Prof. Heiko Balzter is director of the Centre for Landscape and Climate Research at the University of Leicester and Official Development Assistance (ODA) lead in the NERC National Centre for Earth Observation. He holds the Royal Society Wolfson Research Merit Award and the Royal Geographical Society's Cuthbert Peek Award. His research focuses on satellite Earth observations of landscapes and climate.

Can you describe your work related to climate change?

As director of the Centre for Landscape and Climate Research at the University of Leicester and ODA[40] lead in the NERC National Centre for Earth Observation, I lead a research program into Earth observation applications to monitor landscape changes and climate change impacts.

What concerns you the most about climate change in your country?

The UK has experienced highly variable and quite erratic rainfall, persistent drought, long cold winters and heat waves over the past decade. These affect land management, agriculture and the water cycle. Some species are declining while new invasive species are found here now. At the same time, water companies are struggling to adapt to climate change to meet water demands of the future. Harvests are becoming uncertain and I am concerned about future food security in the UK, particularly in the context of isolationist policies around Brexit.

What concerns you the most about climate change globally?

I am concerned about the submergence of small island states and the threat to coastal populations, the catastrophic failure of agricultural crops in some years in low-income countries and the associated instability and global insecurity, as hunger and poverty often lead to armed violence and terrorism.

Can you describe a specific sustainability project in your area or country that is contributing to a "greener world" and lower carbon footprint?

The Kettering Energy Park will generate enough electricity to satisfy nearly all of the areas' current and future requirements. Phase one, known as the Burton Wold Wind Extension, will see the installation of nine high efficiency General Electric turbines that will produce enough power for 11,000 homes. Phase two will see the installation of a solar array.

The Energy Park will act as an enabling development for existing consented schemes within Kettering that have to meet renewable energy targets (typically planning consents now require increasing amounts of energy to come from renewable sources).

The Energy Park provides a direct benefit to the community in that it provides a resilient and robust supply of electricity for all domestic, commercial and municipal purposes.

It also provides a dedicated area for high energy users and consumers.

See http://www.ketteringenergypark.co.uk/.

What steps could your government take to mitigate the growing threats from climate change?

The UK could develop a pathway to a zero-carbon economy. This would entail research into sustainable transport (including cheaper and more reliable public transport to get people out of their cars), a policy to stop fracking and leave hydrocarbons underground rather than burning them and release more GHGs into the atmosphere, and to re-establish support for renewable energies such as tidal, solar and wind. I exclude nuclear power from the range of solutions, because of the long-term costs of storing nuclear waste and the high risk of operating nuclear installations.

What is your reaction to the UNIPCC's accepted calculation by Prof. Myers of a minimum of 200 million climate change refugees by 2050 and the world's readiness to cope with this humanitarian and environmental challenge.

I believe that this estimate could be realistic or possibly an understatement. Already, millions of people are migrating worldwide to escape poverty, war and the loss of their livelihoods. This will be greatly exacerbated by climate change once the impacts are being felt to their full extent. Africa is hard-hit by climate change because of the intersection of poverty and vulnerability to climate change impacts.

Can you describe a particular area of climate change that you are concentrating upon with research or a grassroots development project?

A large part of my current research concentrates on using satellite Earth observation to deliver information on global forest resources. These data support the implementation and monitoring of the UNs' REDD+ initiative which seeks to reduce GHG emissions from deforestation and forest degradation by rewarding forested nations for protecting their forests.

See for example, Lynch, J., Maslin, M., Balzter, H. and Sweeting, M. (2013): Sustainability: Choose satellites to monitor deforestation, Nature 496, 293–294. http://hdl.handle.net/2381/28888.

Interview

Ida Maria Pinnerod is the Mayor of Bodø, Norway, the first zero emissions town in the world and a leader in sustainability policy. By 2050, a forecasted 70% of global population will reside in cities. Moreover, 70% of climate gas emissions come from cities, and cities account for 60% of global energy consumption. Bodø is the pioneer in sustainable zero-emission city design.

Can you describe the climate change threats facing Bodø and Norway?

In Nordland there has been an increase of almost 20% in the annual rainfall in the last century. The rainfall is highest in winter (24%) and at lowest in summer (8%). For Bodø municipality, observations show an increase of approximately 15%. This corresponds to an increase of 150 mm, which corresponds to a normal October month. Generally, we will see an increase in precipitation up to 2100. The greatest uncertainty is related to how general weather patterns will respond to the changes that are taking place, especially in relation to the disappearance of ice cover in the Arctic. This results in the fact that, even if there is more precipitation, longer periods of drought may also be experienced in the future. For Nordland, an annual growth rate of 5% is expected in a low-emission scenario, while the increase is expected to exceed 40% in a high-emission scenario. This greatest increase in rainfall will occur in summer and autumn. The precipitation increase observed over the last 30 years is close to the high-emission scenario.

Bodø municipality has a large watercourse where the Norwegian Water Resources and Energy Authority (NVE) has prepared a flood map—this is Lakselva in Misvær. In addition, there are flood problems related to Futelva and Bodøelva. Other major watercourses are not considered as problematic in relation to floods as no infrastructure or buildings are closely linked to these watercourses. The Bodø peninsula consists of a number of smaller waterways, where floods can create challenges, but this problem is considered non-existent today. However, in the future, higher precipitation levels will actualize problems with flooding in the Bodø region.

Drainage and flooding: In the case of large amounts of precipitation in a short period of time, problems may be encountered with getting water away quickly. Right now, this is not a big problem in Bodø. The topography in developed areas is such that the water relatively quickly finds its way to the sea. The problem will nevertheless increase in the future with increasing episodes of rainfall. With the good starting point that Bodø municipality has in this area, this problem will be minimized through good development planning.

Shredding: Shreds occur in many different forms. The Norwegian Geotechnical Institute has divided them into three main types, snow, loose and stone rocks. Furthermore, the slopes are divided into ten different subcategories. All categories are more or less relevant for Bodø municipality and must therefore be considered. An increase in all types of slopes is expected in the future due to increased precipitation and increased frost blasting. With higher temperatures, the area exposed to frost blasting will move upwards in height. This could cause more rock jumping, and rocks in areas where they are not common today.

Mountain slopes in Bodø municipality are not mapped. The NVE will start mapping them in 2016. It could be potential mountain cliffs in Bodø municipality or neighboring municipalities that could damage the Bodø community. In a warmer climate, permafrost will melt and increase the potential for rock and mountain cliffs.

The risk of loose mass cuts will increase with increased intensity of rainfall, in particular there will be an increasing frequency of landslides during the summer.

Rapid temperature fluctuations combined with more rainfall will increase the risk of snow and southern sprains.

Sea level rise: The increase in sea level is mainly due to two things. Heating water causes the water to expand. The warming of the oceans causes water levels to rise. Melting of glaciers leads to increased drainage to the sea and thus increasing water levels. In Norway, sea level rise is halted by landing. We have therefore not experienced the major effects of sea level rise as the increase in water levels has largely been lifted by landing. In years to come it is expected that the water level will rise faster than the landing and we will get sea level rise.

When sea level rise is calculated over time, there are some uncertainties. For planning purposes, it is important to add the highest estimate as a basis for future marine development. In Bodø, sea level is expected to rise by 27 cm by 2050 and by 89 cm by the year 2100 if we add the highest figures.

Saltstraumen has a dampening effect on the tide so that the tidal rash is 60 cm lower in the Skjerstadfjord than in Bodø municipality, but the sea level rise will nevertheless be the same, but the flood will be 60 cm lower in Saltstraumen than outside.

Wind: Wind observations do not show any trend for wind turbine change. Generally, wind observations are associated with uncertainty as change in equipment over time and change in location of measurement stations is more effective in measuring wind results than other measurable parameters. Generally, an increase in storm activity in the Norwegian Sea is expected in a warmer climate, but there is no expected increase in observed wind power. It is therefore common to assume that in the future there will be more frequent extremes due to wind, but the extremes do not get stronger than we observe today.

Business development: Different industries will experience different effects of future climate change. Industries that are heavily dependent on natural resources, such as agriculture, forestry, fisheries and more, are most vulnerable and may experience both the negative and positive effects of a changing climate. Tourism will also face challenges with a changing climate. The report "Analysis of expected climate change in Nordland" prepared by Nordlandsforskning for Nordland county municipality in

2010 summarizes relevant research on the effects climate change will have on the various industries.

Biological diversity: Primary industries rely on bio-diversity and resource access. A warmer climate will affect bio-diversity. The report "Effects on ecosystems and biodiversity—Climate change in the Norwegian Arctic" published by the Norwegian Polar Institute provides a summary of the effects climate change will have on bio-diversity in northern Norway.

In a warmer climate, many species will move their range higher up and further north to be in a climate that they have adapted. The species that we lose first are those species that have no place to emigrate. This largely applies to species that are adapted to a life in the high mountains.

Increased temperatures will provide an earlier start to the growing season, a longer growing season and increased primary production. Both pine forests and birch forests will increase their distribution to the north and at higher altitude. For plant breeders, increased primary production and milder winters will increase stocks. Particular species diversity and stocks of deer and insects are expected to increase. The migration and hunting period of birds will be shifted in order to match the modified growth of insects. Increased temperature and increased species diversity of hosts will provide a greater variety of parasites. Problems with mosquitoes, biting insects and ticks are expected to increase.

Agriculture: A warmer climate will produce shorter winters and a longer growing season. Longer growing times will give greater crops yields and opportunity for more harvests. Increased temperature can also provide the basis for cultivating more heat-intensive crops as well as grain production.

Growing season is defined based on the number of days a year with a temperature of above 5 °C. There is uncertainty as to how far the growing season will be expanded, depending on the studies that are done. A conservative approach assumes 30 to 40 more days in the growing season in 2050 compared with today. Nordland from Salten and the north is the area in Norway where the seasonal extension is expected to be greatest. Further towards the year 2100, the growing season is expected to be up to three months longer inland, and somewhat shorter along the coast.

Changes in the climate are not only positive for agriculture, one can expect greater opportunities for erosion and drainage, especially in the autumn. An increase in plant diseases and pests may be expected, but the problems are not expected to be as great in northern Norway as in the rest of the country. Today, the fjords are more exposed to ice fires than coastal and inland areas. In a warmer climate, it is expected that this problem will largely be an inland phenomenon. This will therefore be a positive effect for agriculture in Bodø municipality.

In sum, it is expected that climate change could have a positive impact on agriculture in Bodø and Salten.

Forestry: As for agriculture, forestry will also experience the positive and negative effects of climate change. With a longer growing season, a greater growth and production potential for wood production and bio-energy is expected. An increased growing period will also increase the potential for carbon storage. Productive areas for forest production are expected to increase sharply, perhaps as much as 30% due to a warmer climate. This is mainly due to the fact that forest boundaries are moved to higher altitudes. With a temperature rise of 2 °C it is expected that the forest boundary will move up by about 300 m. This means that the entire height zone of between 300 and 600 m is expected to be below the forest boundary in the future. In a warmer climate, the forest could be exposed to more pests and diseases with pests moving rapidly northwards. In addition, forests will be more vulnerable to forest fires during longer and warmer droughts due to climate change.

Overall, however, it is expected that forestry in Bodø and Salten will experience a positive effect from climate change.

Fisheries and aquaculture: Fishing is one of Norway's most important industries and is also an important industry for Bodø. Nearly 80% of the heat inflicted on the global climate system is absorbed by the ocean. The fish in the sea are affected both directly and indirectly by a warmer sea. The fish energy turnover increases in warmer water so that the growth of fish generally increases. Temperature also helps to change sex maturity and egg production. Indirectly, climate change affects the amount of fish feed (plant and animal plankton and smaller fish), which in turn affects the spread and growth of the fish.

Nordlandforskning has reviewed relevant research in fisheries and climate change through the report "Analysis of expected climate change in Nordland." It concludes that a 1–2 °C rise in ocean temperature will likely increase recruitment of Norwegian springy herring and northeast Arctic cod. Both are commercially important fish stocks. At the same time, fishermen like mackerel, and mackerel will increase their northern distribution.

The aquaculture industry is sensitive to temperature changes. Salmon thrives best at temperatures of between 5 and 20 °C. At higher temperatures, salmon will experience high mortality. Due to climate change, locations in southern Norway/western Norway could become unsuitable for farming due to high temperatures in the water in late summer. In Nordland, however, higher sea temperatures will be positive for the aquaculture industry as higher temperatures provide higher growth rates and productivity. Studies show that profitability in the aquaculture industry in Nordland will increase by 25% per degree Celsius of temperature increase as long as the temperature increase does not exceed 2.5 °C. Climate change is expected to have a positive effect on fisheries and aquaculture.

Reindeer husbandry: Most studies on reindeer husbandry and climate change have been carried out in Finnmark, where there are summer pastures on the coast and winter pastures on the shore. Shorter winters on the coast will generally lead to increased survival for the reindeer. Earlier, we are expected to increase calf weight, but increased insect pests in the summer could increase calf mortality. Increased temperature can also increase the risk of parasites and diseases.

Longer droughts will become more common in the future and will adversely affect reindeer husbandry. In all, it is assumed that reindeer husbandry will experience adverse effects of climate change.

Tourism: The impact on tourism is difficult to predict, as it depends on how the climate is developing, but also what policy measures are taken to reduce GHG emissions. A direct effect of climate change is shorter winters and less snow. This will have a direct effect on the ski season. In areas that are currently marginal and dependent on the production of artificial snow, there will be no basis for skiing in the future when the temperature is on the plus side most of the winter. Changes in tax regimes on means of transport to reduce global CO_2 emissions could have adverse effects on tourism in northern Norway.

Can you describe the plan to make Bodø a zero emissions town?

National and international agreements for Norway assumed that the first plan for zero emissions was made for Bodø municipality in 2008. The plan was ambitious and aimed at Bodø municipality to be a zero emissions city in 2020. Following various political decisions in Bodø city council, a new climate plan was launched in 2012. The plan was adopted in autumn 2014. One of the measures was to develop the city into a zero emissions city. Major upheavals in the Bodø community when the battlefield base was shut down meant that a new plan was needed. It was proposed that the airport be moved south to the Bodø Peninsula leaving a large area open for urban development, a zero discharge city.

The municipalities play an important role in reducing Norway's GHG emissions and for converting energy into environmentally friendly energy use. Climate and energy plans are an important local policy management tool for this.

A municipality has many roles in local climate and energy work. It manages legislation and plans the use of land and the transport system. It is an important source of knowledge and can act as a driver for further development of sustainable communities.

A municipality has a wide range of instruments to reduce GHG emissions, which are linked to area planning, agricultural management, waste management and energy use. Furthermore, a municipality can influence companies and events to think about climate and the environment through the management of economic instruments and procurement schemes. Through the Government Plan for Climate and Energy Plans, municipalities are required to take note of climate concerns and to convert to more environmentally friendly energy use. They implement measures to reduce GHG emissions and ensure more efficient energy use and environmentally friendly energy conversion in accordance with this policy or in their own municipality plan.

Regulations on energy exploration require local area consultant to prepare local energy research. For Bodø municipality, Nordlandsnett AS is preparing this investigation. The investigation will provide information on local energy supplies, stationary energy use and alternatives in this area. The local energy survey is an important source of information and a basis document for municipal climate and energy planning. Proposals have been made for the system of local energy research to be settled. This will be clarified during 2014/2015.

Bodø municipality is also assigned tasks in the area of climate and energy through the Act on Municipal Emergency Planning, Civil Protection and Civil Defense Act (Civil Protection Act) with corresponding regulations on municipal emergency preparedness. Climate change adaption is a field of expertise that belongs in both the municipal subdivision for climate and energy and in the municipality's comprehensive risk and vulnerability analysis.

The framework for international climate cooperation is the "UN Framework Convention on Climate Change" (Climate Change Convention). The convention was adopted in 1992 and has been ratified by 195 parties. As a long-term goal, the Climate Convention aims to stabilize the concentration of GHGs in the atmosphere at a level that prevents a dangerous and negative man-made impact on the climate system. The convention does not contain quantified emission commitments for individual parties, but states that industrialized countries must be in the forefront of combating climate change and their negative effects, and demonstrate this through national instruments and measures.

The Kyoto Agreement was finalized and adopted at the December meeting of the Kyoto Convention in December 1997. The Kyoto Agreement sets out binding and quantified limits on GHG emissions. In the period 2008 to 2012, Norway was obliged to limit total GHG emissions to 1% above 1990 levels. This means approximately 52.5 million metric tons of CO_2 equivalents. Emissions over this figure would have to be offset by the purchase of climate quotas abroad. The Kyoto Agreement's second commitment period runs from 2013 to 2020. Norway has committed itself to reducing emissions to 84% of 1990 levels. The challenge with the Kyoto Agreement is that some major emissions countries fall outside the system either because they have not ratified the agreement or are exempted from emission reductions since they are covered by the U-zone area of the agreement. This means that only 15% of total global emissions are covered by the Kyoto Agreement. On October 24, 2014, the European Council adopted binding targets for a reduction of GHG emissions by at least 40% by 2030 compared with 1999. At the EU summit, a target of a renewable energy share of at least 27% by 2030 was also agreed. The agreed target is binding at an EU level but without national binding targets, that is, it gives the countries negotiating opportunities in between.

In addition, EU heads of state and government set targets for improving energy efficiency by at least 27% and aiming at reaching 15% cross-border electricity transmission capacity, ensuring that no member states are isolated from access to energy. The outcome of negotiations in the EU formed the basis for further negotiations for the climate summit in Paris in 2015. Norway does not automatically implement the EU's goals so far, national rules as before apply.

The national guidelines include:

1. Report. 21 (2011–2012) Norwegian climate policy;
2. NOU 2010: 10 adaptation to a changing climate;
3. Report. St. 33 (2012–2013) Climate adaptation in Norway;
4. national expectations for regional and municipal planning (T-1497, 201);
5. Renewable Directive (2009/28/EC, entered into force on 20.12.11);
6. St. meld. 39 (2008–2009) Climate challenges—agriculture a part of the solution;
7. state plan for coordinated housing, area and transport planning.

In Report. 21 (2011–2012) the government has stated that the Norwegian climate policy sets the following long-term goals:

1. Within the Kyoto Protocol's first commitment period, Norway will over-reach the Kyoto commitment by percentage points.

2. By 2020, Norway will undertake a commitment to cut global emissions of GHGs, equivalent to 30% of Norway's emissions in 1990.
3. Norway must be carbon neutral in 2050.

The government's 2020 target includes both emission reductions in Norway, including carbon capture in forests, and Norway's contribution to emission reductions in other countries. The government's goal is to cut about two-thirds of domestic emissions. This corresponds to a reduction of 12–14 million metric tons of CO_2 equivalents when crediting or forest admissions are included. Net recordings in forests, according to Report No. 21 (2011–2012) Norwegian climate policy, in recent years have been 27–36 million metric tons of CO_2 equivalents annually. Through the Kyoto Protocol, Norway has not been allowed to refund more than 3% of total emissions as forests. It will be about 1.5 million metric tons of CO_2 equivalents.

Norway is responsible for reducing global emissions of GHGs, equivalent to 100% of its own emissions by 2050. If a global and ambitious climate agreement is concluded, where other industrialized countries also assume major commitments, Norway wants a binding target of carbon neutrality by 2030. Conditions that may affect these goals are developments in international carbon prices, increased costs for implementing climate measures in Norway, population growth, increased GDP, increased petroleum emissions and delayed development of climate friendly technology. A reduction in international carbon prices will make it cheaper to buy quotas than implement climate measures. Increased labor migration will contribute to an increase in the population of approximately 500,000 people over the population base underlying climate change. GDP for the mainland economy is expected to be 10% higher in 2020 than that of climate change.

All of these factors will contribute to increased emissions. A continuation of today's instruments will therefore lead to an increase of 49.8 million metric tons of CO_2 equivalents in 1990 to 56.9 million metric tons by 2020. The government therefore proposes to strengthen the use of instruments in national climate policy. This is due to increased production of renewable energy, the development of new climate friendly technology in Norway and the use of technology developed in other countries.

Describe some exciting features about the zero emissions plan for Bodø.

Bodø is the city in Norway with the most electric cars in relation to its population. All electricity used in industry and in households is green renewable energy from hydropower plants.

Projects with driverless electric buses, garbage trucks, major investments for pedestrians and cyclists have been launched in Bypakke Bodø.

A separate research project on the use of solar cells on roofs has been started.

The potential for using seawater as heating through heat exchange will reduce energy consumption by up to 20%.

Forests play an important role in binding and storing CO_2. On a national basis, between 27 and 36 million metric tons of CO_2 is absorbed per year due to the growth of forests. In Bodø, it is estimated that 100,000 tonnes of CO_2 will be bound each year due to the growth of forests. The growth of forests thus offsets a significant amount of internal CO_2 emissions within the boundaries of Bodø municipality. The standing volume of forest in Bodø municipality is intended to store approximately 2.7 million metric tons of CO_2.

To act as a carbon layer the forest must be properly sealed. Forest trees must be chopped down and used before they start to rot, thus releasing CO_2 and methane. In addition, it is important to add new forest through ordinary forest planting and the creation of climate forest areas.

Increased use of wood in building materials causes carbon to be stored while trees replace more climate-challenging and energy-intensive building materials.

Bypakke Bodø has a number of measures aimed at soft road users and public transport. On June 20, 2014, the government sent Bill 131 (2013–2014) to the Storting for further processing. The Bypakke has a total financial framework of NOK 2820 million (2014). This involves a new highway 80 from Hunstadmoen to the Thallekrysset, estimated to cost NOK 2080 million. Furthermore, measures have been taken for pedestrians and cyclists (NOK 500 million) and public transport (NOK 210 million). On the collective side there will also be a county municipality investment of NOK 59 million in the period 2014–2021. In addition, the development of train stops in Tverlandet is also included.

A realization of the city package is the most important step in reducing GHG emissions from traffic. It is therefore important to ensure that proposed measures in the package are implemented, including a road safety plan and a cycle plan. The public transport system in Bodø municipality was restrained on October 1, 2012. The conversion has been a success and has resulted in an increase of 18% in the number of bus passengers—250,000 more trips in 2013 than in 2012. The potential for further growth is large if we compare the number of bus travelers in Bodø with other cities such as Tromsø. Measures through the city package must therefore make it more attractive and easier to travel by bus than by car.

In Bodø, Fauske and Saltdal, Saltenpendelen is an important public offer. Saltenpendelen currently has stops in Bodø city center, Mørkved, Valnesfjord, Fauske and Rognan. The crossing stop opened in July 2015. Furthermore, the government has granted NOK 34 million in compensatory funds for increased employer's contribution to railway development in Salten. The allocations will be used to extend the crossing track at Oteråga from 290 to 600 m. The project also includes construction of a platform to serve travelers who are affiliated with the Reitan defense facility.

Today there are a number of environmental certification schemes where the lighthouse is the most well known. By fulfilling the requirements for environmental lighthouses, businesses commit themselves to actively following up environmental measures, and to having systems for monitoring environmental work including the working environment. Bodø municipality is a certification officer for those companies that wish to be certified. Bodø municipality has 86 companies/businesses that are certified as environmental lighthouses, including municipalities such as the City Hall and the Municipal Office. Bodø municipality as a purchaser has the opportunity to require suppliers to be environmentally certified through the environmental lighthouse scheme or equivalent schemes (ISO 14001). An eco-approved event is an award for organizers who will actively work to reduce the environmental impact of an event. An organizer must meet certain criteria in several areas to receive the award. Any type of event that can meet the criteria can be approved.

How are citizens and community groups participating to make Bodø a zero emissions town?

Processes with the involvement of the city's residents, business, institutions and associations are being put in place.

What is your reaction to the forecast by Prof. Myers and the IPCC of a minimum of 200 million climate change refugees by 2050 and the world's readiness to cope with this humanitarian and environmental challenge?

We have assumed in our planning that our city will be a zero emission city and we will have a sustainable business community based on renewable natural assets. Fisheries and agriculture will in future find significant value in settlement in our region. The climate can be changed in such a way that we either get more production or we need to change our food production. If we have to look at Norway and northern Norway as a winner in climate change around the globe, this can mean a large relocation of people. Today, we cannot handle this for cultural reasons.

Can you discuss reactions from communities in Norway and other countries to the Bodø plan?

Reactions from other communities in Norway have been very positive. The biggest reactions, however, have come from other influential countries, such as China, the UK and Germany. International support and backing in our vision for the future have propelled both the development of and belief in the Bodø plan. These reactions showcase international awareness when it comes to facing climate challenges, and how we need to deal with them.

Notes

1. EJF (2009) No Place Like Home—Where next for climate refugees? Environmental Justice Foundation: London.
2. IPCC, 2014: Summary for policymakers. In: Climate Change 2014: Impacts, Adaptation, and Vulnerability. Part A: Global and Sectoral Aspects. Contribution of Working Group II to the Fifth Assessment Report of the Intergovernmental Panel on Climate Change [Field, C.B., V.R. Barros, D.J. Dokken, K.J. Mach, M.D. Mastrandrea, T.E. Bilir, M. Chatterjee, K.L. Ebi, Y.O. Estrada, R.C. Genova, B. Girma,

E.S. Kissel, A.N. Levy, S. MacCracken, P.R. Mastrandrea, and L.L. White (eds.)]. Cambridge University Press, Cambridge, United Kingdom and New York, NY, USA, pp. 1–32.
3. ASC (2016) UK Climate Change Risk Assessment 2017 Synthesis Report: priorities for the next five years. Adaptation Sub-Committee of the Committee on Climate Change, London.
4. EA (2009a) Flooding in England: A National Assessment of Flood Risk. Environment Agency, UK.
5. Climate Change and Health Country Profile—2015 United Kingdom, WHO and UNFCCC, 2015.
6. Met Office. 2016. Winter 2015/2016. Met Office. http://www.metoce.gov.uk/climate/uk/summaries/2016/winter.
7. Climate Prediction.net. 2016. 2015 December Extreme weather in the UK. Climate Prediction. http://www.climateprediction.net/weatherathome/2015-december-extreme-weather-in-the-uk/.
8. ASC (2016) UK Climate Change Risk Assessment 2017 Synthesis Report: priorities for the next five years. Adaptation Sub-Committee of the Committee on Climate Change, London.
9. "Temperature effects of climate change on human health", Shakoor Hajat, London School of Hygiene and Tropical Medicine Sotiris Vardoulakis, Health Protection Agency Clare Heaviside, Health Protection Agency Bernd Eggen, Health Protection Agency in "Health Effects of Climate Change in the UK 2012 Current evidence, recommendations and research gaps" Sotiris Vardoulakis and Clare Heaviside (Editors) Health Protection Agency, 2012.
10. Hanson et al. (2010); Richards and Nicholls (2009), both in: Met Office (2011).
11. Pardaens et al. (2011), in: Huntingford et al. (2014).
12. Kovats, R.S., R. Valentini, L.M. Bouwer, E. Georgopoulou, D. Jacob, E. Martin, M. Rounsevell, and J.-F. Soussana, 2014: Europe. In: Climate Change 2014: Impacts, Adaptation, and Vulnerability. Part B: Regional Aspects. Contribution of Working Group II to the Fifth Assessment Report of the Intergovernmental Panel on Climate Change [Barros, V.R., C.B. Field, D.J. Dokken, M.D. Mastrandrea, K.J. Mach, T.E. Bilir, M. Chatterjee, K.L. Ebi, Y.O. Estrada, R.C. Genova, B. Girma, E.S. Kissel, A.N. Levy, S. MacCracken, P.R. Mastrandrea, and L.L. White (eds.)]. Cambridge University Press, Cambridge, United Kingdom and New York, NY, USA, pp. 1267–1326.

13. European Commission, 2009, Policy Research Corporation, Country Assessment Italy, 2009. https://ec.europa.eu/maritimeaffairs/sites/.../files/docs/.../italy_climate_change_en.pdf.
14. Ministero dell'Ambiente (2000), in: Ministry for the Environment, Land and Sea of Italy (2007).
15. Ibid.
16. Regional climate change and adaptation—The Alps facing the challenge of changing water resources. EEA Report No 8/2009. EEA, Copenhagen.
17. European Environment Agency, "The vulnerability of the Alps to climate change—Climate change and impacts—Future threats—The need to adapt" 2010 and EEA, 2009.
18. "Ensemble projections of future streamflow droughts in Europe" G. Forzieri, L. Feyen, R. Rojas, M. Flörke, F. Wimmer, and A. Bianchi, Climate Risk Management Unit, Institute for Environment and Sustainability, Joint Research Centre, European Commission, Ispra, Italy, Center for Environmental Systems Research (CESR), University of Kassel, Kassel, Germany, 2014.
19. http://www.dailymail.co.uk/sciencetech/article-2536499/Climate-change-cause-Europe-suffer-80-droughts-claim-scientists.html#ixzz4kaMyXnNa.
20. Istituto Superiore per la Protezione e la Ricerca Ambientale, Climate indicators in Italy 2013—edition IX, Climate indicators in Italy 2013—edition IX http://www.isprambiente.gov.it/en/publications/state-of-the-environment/climate-indicators-in-italy-2013-edition-ixvia@ISPRA_Press2013.
21. The European Environment Agency (EEA, 2015).
22. Kovats, R.S., R. Valentini, L.M. Bouwer, E. Georgopoulou, D. Jacob, E. Martin, M. Rounsevell, and J.-F. Soussana, 2014: Europe. In: Climate Change 2014: Impacts, Adaptation, and Vulnerability. Part B: Regional Aspects. Contribution of Working Group II to the Fifth Assessment Report of the Intergovernmental Panel on Climate Change [Barros, V.R., C.B. Field, D.J. Dokken, M.D. Mastrandrea, K.J. Mach, T.E. Bilir, M. Chatterjee, K.L. Ebi, Y.O. Estrada, R.C. Genova, B. Girma, E.S. Kissel, A.N. Levy, S. MacCracken, P.R. Mastrandrea, and L.L. White (eds.)]. Cambridge University Press, Cambridge, United Kingdom and New York, NY, USA, pp. 1267–1326.
23. Rosenzweig C., Tubiello F.N. (1997) "Impacts of global climate change on Mediterranean Agriculture, current methodologies and future

directions: an introductory essay" Mitigation and Adaptation Strategies for Global Change 1, 219–232.
24. "Climate change effects and Agriculture in Italy: a stochastic frontier analysis at regional level" Sabrina Auci and Donatella Vignani. Preliminary Draft January, 2014.
25. "Italy Climate Change, Country overview and assessment, The economics of climate change adaptation in EU coastal areas" Policy Research Corporation, 2014.
26. "The Environmental, Economic and Social Impacts of Climate Change in Greece", Bank of Greece, June, 2011.
27. Hellenic Republic, Ministry for the Environment, Physical Planning and Public Works of Greece (2006).
28. Bank of Greece (2011), in: Shoukri and Zachariadis (2012).
29. "The Environmental, Economic and Social Impacts of Climate Change in Greece", Bank of Greece, June, 2011.
30. "The Environmental, Economic and Social Impacts of Climate Change in Greece", Bank of Greece, June, 2011.
31. "The Environmental, Economic and Social Impacts of Climate Change in Greece" Christos Zerefos, 2012.
32. European Environment Agency, "Climate change poses increasingly severe risks for ecosystems, human health and the economy in Europe" January 25, 2017.
33. "The Environmental, Economic and Social Impacts of Climate Change in Greece", Bank of Greece, June, 2011.
34. "Future Projections of Heat Waves in Greece. Extreme or Common Events?" P.T. Nastos and J. Kapsomenakis in C.G. Helmis and P.T. Nastos (eds.), Advances in Meteorology, Climatology and Atmospheric Physics, Springer Atmospheric Sciences, https://doi.org/10.1007/978-3-642-29172-2_90, # Springer-Verlag Berlin Heidelberg 2012.
35. Kovats S. et al. Heat waves and human health. In: Menne B., Ebi K.L., eds. Climate change and adaptation strategies for human health. Heidelberg, Springer Verlag, 2006: 63–90.
36. The PESETA project [web site]. Brussels, European Communities, 2008 http://peseta.jrc.es.
37. The Government Gazette, "Climate Change Issues and the Need for Growth in Greece" by Yannis Maniatis, Minister of Environment, Energy, and Climate Change, April 15, 2014.

38. "The environmental, economic and social impacts of climate change in Greece" Christos Zerefos, 2012.
39. Climate Change Post, Greece, July 26, 2017. https://www.climatechange-post.com/greece/fresh-water-resources/ 2014.
40. Official Development Assistance.

Conclusion: Climate Change Projections to 2100

Some 25 centuries ago, the famed Athenian statesman Pericles wisely observed that "where there is no vision, the people shall perish." Today, a lack of vision with respect to climate change adaption and mitigation will lead to populations and nations that indeed perish from flooding, drought, health crises and environmental destruction. The signs are clear and undeniable in all parts of the world where weather phenomena triggered by climate change are becoming increasingly evident and dangerous.

Climate projections for the year 2100 are daunting:

1. The planet will heat up by an estimated 2–3 °C.
2. There will be major flooding and climate refugee movements in the millions from the cities of New York, Miami, Houston and Los Angeles. Hurricane Irma in 2017 tragically demonstrated the destructive power and pattern of hurricane storms in Texas and Florida and flood devastation.
3. Global warming creates more moisture in the environment which enhances and propels the force of hurricanes.

4. As global warming contributes to elevated moisture in the atmosphere it also triggers stronger monsoon rainfall events as demonstrated in India, Bangladesh and Southeast Asia in 2017.
5. We will see severe drought in northern China and the desertification of hundreds of towns and communities.
6. There will be major flooding of coastal metropolis cities including Shanghai, London, Kolkata, Manila, Bangkok, Venice and Mombasa.
7. There will be critical flooding and partial submergence of small island nations including the Maldives and Marshall Islands.
8. We will see an estimated climate refugee population of 200 million by at least 2050 and upwards of 500 million according to some experts.
9. Severe water shortages and scarcity triggered by climate change factors including drought will devastate scores of countries including India, China, Vietnam, Greece, Italy, China and many impoverished nations in central and southern Africa.
10. Sea level will rise by nearly 2 m if GHG emissions continue at their present levels.
11. We will see major food shortages and a dramatic spike in food prices because of climate change induced damage to millions of acres of productive farmland.
12. A dramatic spread of diseases including malaria will be exacerbated by climate change.
13. The capacity of impoverished developing nations to utilize adaption and mitigation strategies will be severely limited. Moreover, the ability of these countries to address and manage major disease symptoms arising from climate change will be profoundly compromised.
14. There are projections that the flow of climate refuges will lead to militarized borders in numerous highly developed industrialized states.
15. An estimated 90% of Arctic ice will disappear by 2100 with attendant sea level rise and coastal flooding.
16. Acidification of the oceans will severely deplete fish stocks while eliminating certain marine species.
17. In North America in 2017, the Canadian province of British Columbia experienced the worst forest fires in decades. Projected

increases in drought episodes combined with elevated temperature projections will exacerbate the forest fire crisis.
18. In India, dramatically falling water tables and severe water scarcity will become critical in the coming decades for a country that is projected to have the worlds' largest population of 1.6 billion by 2050. By 2100, this crisis will lead to health, economic and food shortage devastation.
19. China is spending USD 62 billion to divert water to the north to forestall critical water shortages and drought that is already prevalent. The South-to-North Water Diversion Project constitutes the largest water diversion project ever launched in history. Yet climate change may hinder some of the expected benefits. By 2100, the combined climate phenomena of drought in the north and flooding in the south will create crisis conditions in the country and the potential for severe political upheaval.
20. Drought, sea level rise, coastal flooding and extreme heat episodes will be common challenges in Greece and exacerbate conditions facing a population that is battling severe economic stress.
21. In Africa, many nations will be devastated by rising drought episodes, critical water shortages and escalating food prices. Heat stress and heat related illnesses will rise dramatically.
22. The first six months of 2017 was the worlds' second-warmest on record after 2016. Heat waves in North Africa and the Middle East threaten to become a continuing pattern.

According to National Oceanic and Atmospheric Administration (NOAA) data, the worlds' five warmest years were recorded since 2010:

1. 2016 (1.93 °C above average)
2. 2017 (1.64 °C above average)
3. 2015 (1.55 °C above average)
4. 2010 (1.39 °C above average)
5. 2014 (1.30 °C above average)

Africa is projected to be the continent facing the most severe climate change burdens. Numerous economic and health burdens will confront populations in several countries. Malaria rates for example are expected to increase dramatically. In 2015, according to the World Health Organization, Africa accounted for 90% of global malaria cases.

The data on climate change is well documented and the dangers to populations are well illustrated. The call to action on the need for climate change mitigation and adaption has been sounded by a multitude of governments, climate scientists, engaged groups, citizens and international forums such as the IPCC and Paris Agreement. The world must respond with policies and funding that is bold, urgent and also inclusive to uplift the hardest hit societies in the developing world who, experts agree, will bear the brunt of climate change destruction. "It is better to light a candle than curse the darkness," Eleanor Roosevelt stated. It is time for the world community working with collective urgency to act, innovate and legislate to address the perilous path of climate change, which transcends borders and confronts all humankind.

Appendix A: The Paris Agreement

UNITED NATIONS FRAMEWORK CONVENTION ON CLIMATE CHANGE
UNITED NATIONS 1992 FCCC/INFORMAL/84 GE.05-62220 (E) 200705
UNITED NATIONS FRAMEWORK CONVENTION ON CLIMATE CHANGE

The Parties to this Convention, Acknowledging that change in the Earth's climate and its adverse effects are a common concern of humankind,

Concerned that human activities have been substantially increasing the atmospheric concentrations of greenhouse gases, that these increases enhance the natural greenhouse effect, and that this will result on average in an additional warming of the Earth's surface and atmosphere and may adversely affect natural ecosystems and humankind,

Noting that the largest share of historical and current global emissions of greenhouse gases has originated in developed countries, that per capita emissions in developing countries are still relatively low and that the share of global emissions originating in developing countries will grow to meet their social and development needs,

Aware of the role and importance in terrestrial and marine ecosystems of sinks and reservoirs of greenhouse gases,

Noting that there are many uncertainties in predictions of climate change, particularly with regard to the timing, magnitude and regional patterns thereof,

Acknowledging that the global nature of climate change calls for the widest possible cooperation by all countries and their participation in an effective and appropriate international response, in accordance with their common but differentiated responsibilities and respective capabilities and their social and economic conditions,

Recalling the pertinent provisions of the Declaration of the United Nations Conference on the Human Environment, adopted at Stockholm on 16 June 1972,

Recalling also that States have, in accordance with the Charter of the United Nations and the principles of international law, the sovereign right to exploit their own resources pursuant to their own environmental and developmental policies, and the responsibility to ensure that activities within their jurisdiction or control do not cause damage to the environment of other States or of areas beyond the limits of national jurisdiction,

Reaffirming the principle of sovereignty of States in international cooperation to address climate change,

Recognizing that States should enact effective environmental legislation, that environmental standards, management objectives and priorities should reflect the environmental and developmental context to which they apply, and that standards applied by some countries may be inappropriate and of unwarranted economic and social cost to other countries, in particular developing countries,

Recalling the provisions of General Assembly resolution 44/228 of 22 December 1989 on the United Nations Conference on Environment and Development, and resolutions 43/53 of 6 December 1988, 44/207 of 22 December 1989, 45/212 of 21 December 1990 and 46/169 of 19 December 1991 on protection of global climate for present and future generations of mankind,

Recalling also the provisions of General Assembly resolution 44/206 of 22 December 1989 on the possible adverse effects of sea-level rise on islands and coastal areas, particularly low-lying coastal areas and the pertinent provisions of General Assembly resolution 44/172 of 19 December 1989 on the implementation of the Plan of Action to Combat Desertification, Recalling further the Vienna Convention for the Protection of the Ozone Layer, 1985, and the Montreal Protocol on Substances that Deplete the Ozone Layer, 1987, as adjusted and amended on 29 June 1990,

Noting the Ministerial Declaration of the Second World Climate Conference adopted on 7 November 1990,

Conscious of the valuable analytical work being conducted by many States on climate change and of the important contributions of the World Meteorological Organization, the United Nations Environment Programme and other organs, organizations and bodies of the United Nations system, as well as other international and intergovernmental bodies, to the exchange of results of scientific research and the coordination of research,

Recognizing that steps required to understand and address climate change will be environmentally, socially and economically most effective if they are based on relevant scientific, technical and economic considerations and continually re-evaluated in the light of new findings in these areas,

Recognizing that various actions to address climate change can be justified economically in their own right and can also help in solving other environmental problems,

Recognizing also the need for developed countries to take immediate action in a flexible manner on the basis of clear priorities, as a first step towards comprehensive response strategies at the global, national and, where agreed, regional levels that take into account all greenhouse gases, with due consideration of their relative contributions to the enhancement of the greenhouse effect,

Recognizing further that low-lying and other small island countries, countries with low-lying coastal, arid and semi-arid areas or areas liable to floods, drought and desertification, and developing countries with fragile mountainous ecosystems are particularly vulnerable to the adverse effects of climate change,

Recognizing the special difficulties of those countries, especially developing countries, whose economies are particularly dependent on fossil fuel production, use and exportation, as a consequence of action taken on limiting greenhouse gas emissions,

Affirming that responses to climate change should be coordinated with social and economic development in an integrated manner with a view to avoiding adverse impacts on the latter, taking into full account the legitimate priority needs of developing countries for the achievement of sustained economic growth and the eradication of poverty,

Recognizing that all countries, especially developing countries, need access to resources required to achieve sustainable social and economic development and that, in order for developing countries to progress towards that goal, their energy consumption will need to grow taking into account the possibilities for achieving greater energy efficiency and for controlling greenhouse gas emissions in general, including through the application of new technologies on terms which make such an application economically and socially beneficial,

Determined to protect the climate system for present and future generations,

Have agreed as follows:

Article 1

DEFINITIONS*

For the purposes of this Convention:

1. "Adverse effects of climate change" means changes in the physical environment or biota resulting from climate change which have significant deleterious effects on the composition, resilience or productivity of natural and managed ecosystems or on the operation of socio-economic systems or on human health and welfare.
2. "Climate change" means a change of climate which is attributed directly or indirectly to human activity that alters the composition of the global atmosphere and which is in addition to natural climate variability observed over comparable time periods.
3. "Climate system" means the totality of the atmosphere, hydrosphere, biosphere and geosphere and their interactions.

4. "Emissions" means the release of greenhouse gases and/or their precursors into the atmosphere over a specified area and period of time.
5. "Greenhouse gases" means those gaseous constituents of the atmosphere, both natural and anthropogenic, that absorb and re-emit infrared radiation.
6. "Regional economic integration organization" means an organization constituted by sovereign States of a given region which has competence in respect of matters governed by this Convention or its protocols and has been duly authorized, in accordance with its internal procedures, to sign, ratify, accept, approve or accede to the instruments concerned.

*Titles of articles are included solely to assist the reader.

7. "Reservoir" means a component or components of the climate system where a greenhouse gas or a precursor of a greenhouse gas is stored.
8. "Sink" means any process, activity or mechanism which removes a greenhouse gas, an aerosol or a precursor of a greenhouse gas from the atmosphere.
9. "Source" means any process or activity which releases a greenhouse gas, an aerosol or a precursor of a greenhouse gas into the atmosphere.

Article 2
OBJECTIVE

The ultimate objective of this Convention and any related legal instruments that the Conference of the Parties may adopt is to achieve, in accordance with the relevant provisions of the Convention, stabilization of greenhouse gas concentrations in the atmosphere at a level that would prevent dangerous anthropogenic interference with the climate system. Such a level should be achieved within a time frame sufficient to allow ecosystems to adapt naturally to climate change, to ensure that food production is not threatened and to enable economic development to proceed in a sustainable manner.

Article 3
PRINCIPLES

In their actions to achieve the objective of the Convention and to implement its provisions, the Parties shall be guided, inter alia, by the following:

1. The Parties should protect the climate system for the benefit of present and future generations of humankind, on the basis of equity and in accordance with their common but differentiated responsibilities and respective capabilities. Accordingly, the developed country Parties should take the lead in combating climate change and the adverse effects thereof.
2. The specific needs and special circumstances of developing country Parties, especially those that are particularly vulnerable to the adverse effects of climate change, and of those Parties, especially developing country Parties, that would have to bear a disproportionate or abnormal burden under the Convention, should be given full consideration.
3. The Parties should take precautionary measures to anticipate, prevent or minimize the causes of climate change and mitigate its adverse effects. Where there are threats of serious or irreversible damage, lack of full scientific certainty should not be used as a reason for postponing such measures, taking into account that policies and measures to deal with climate change should be cost-effective so as to ensure global benefits at the lowest possible cost. To achieve this, such policies and measures should take into account different socio-economic contexts, be comprehensive, cover all relevant sources, sinks and reservoirs of greenhouse gases and adaptation, and comprise all economic sectors. Efforts to address climate change may be carried out cooperatively by interested Parties.
4. The Parties have a right to, and should, promote sustainable development. Policies and measures to protect the climate system against human-induced change should be appropriate for the specific conditions of each Party and should be integrated with national development programmes, taking into account that economic development is essential for adopting measures to address climate change.

5. The Parties should cooperate to promote a supportive and open international economic system that would lead to sustainable economic growth and development in all Parties, particularly developing country Parties, thus enabling them better to address the problems of climate change. Measures taken to combat climate change, including unilateral ones, should not constitute a means of arbitrary or unjustifiable discrimination or a disguised restriction on international trade.

Article 4
COMMITMENTS

1. All Parties, taking into account their common but differentiated responsibilities and their specific national and regional development priorities, objectives and circumstances, shall:

    (a) Develop, periodically update, publish and make available to the Conference of the Parties, in accordance with Article 12, national inventories of anthropogenic emissions by sources and removals by sinks of all greenhouse gases not controlled by the Montreal Protocol, using comparable methodologies to be agreed upon by the Conference of the Parties;

    (b) Formulate, implement, publish and regularly update national and, where appropriate, regional programmes containing measures to mitigate climate change by addressing anthropogenic emissions by sources and removals by sinks of all greenhouse gases not controlled by the Montreal Protocol, and measures to facilitate adequate adaptation to climate change;

    (c) Promote and cooperate in the development, application and diffusion, including transfer, of technologies, practices and processes that control, reduce or prevent anthropogenic emissions of greenhouse gases not controlled by the Montreal Protocol in all relevant sectors, including the energy, transport, industry, agriculture, forestry and waste management sectors;

(d) Promote sustainable management, and promote and cooperate in the conservation and enhancement, as appropriate, of sinks and reservoirs of all greenhouse gases not controlled by the Montreal Protocol, including biomass, forests and oceans as well as other terrestrial, coastal and marine ecosystems;
(e) Cooperate in preparing for adaptation to the impacts of climate change; develop and elaborate appropriate and integrated plans for coastal zone management, water resources and agriculture, and for the protection and rehabilitation of areas, particularly in Africa, affected by drought and desertification, as well as floods;
(f) Take climate change considerations into account, to the extent feasible, in their relevant social, economic and environmental policies and actions, and employ appropriate methods, for example impact assessments, formulated and determined nationally, with a view to minimizing adverse effects on the economy, on public health and on the quality of the environment, of projects or measures undertaken by them to mitigate or adapt to climate change;
(g) Promote and cooperate in scientific, technological, technical, socio-economic and other research, systematic observation and development of data archives related to the climate system and intended to further the understanding and to reduce or eliminate the remaining uncertainties regarding the causes, effects, magnitude and timing of climate change and the economic and social consequences of various response strategies;
(h) Promote and cooperate in the full, open and prompt exchange of relevant scientific, technological, technical, socio-economic and legal information related to the climate system and climate change, and to the economic and social consequences of various response strategies;
(i) Promote and cooperate in education, training and public awareness related to climate change and encourage the widest participation in this process, including that of non-governmental organizations; and
(j) Communicate to the Conference of the Parties information related to implementation, in accordance with Article 12.

2. The developed country Parties and other Parties included in Annex I commit themselves specifically as provided for in the following:

   (a) Each of these Parties shall adopt national[1] policies and take corresponding measures on the mitigation of climate change, by limiting its anthropogenic emissions of greenhouse gases and protecting and enhancing its greenhouse gas sinks and reservoirs. These policies and measures will demonstrate that developed countries are taking the lead in modifying longer-term trends in anthropogenic emissions consistent with the objective of the Convention, recognizing that the return by the end of the present decade to earlier levels of anthropogenic emissions of carbon dioxide and other greenhouse gases not controlled by the Montreal Protocol would contribute to such modification, and taking into account the differences in these Parties' starting points and approaches, economic structures and resource bases, the need to maintain strong and sustainable economic growth, available technologies and other individual circumstances, as well as the need for equitable and appropriate contributions by each of these Parties to the global effort regarding that objective. These Parties may implement such policies and measures jointly with other Parties and may assist other Parties in contributing to the achievement of the objective of the Convention and, in particular, that of this subparagraph;

   [1] This includes policies and measures adopted by regional economic integration organizations.

   (b) In order to promote progress to this end, each of these Parties shall communicate, within six months of the entry into force of the Convention for it and periodically thereafter, and in accordance with Article 12, detailed information on its policies and measures referred to in subparagraph (a) above, as well as on its resulting projected anthropogenic emissions by sources and removals by sinks of greenhouse gases not controlled by the Montreal Protocol for the period referred to in subparagraph (a), with the aim of returning individually or jointly to their 1990 levels these anthropogenic emissions of carbon dioxide and other greenhouse gases not controlled by the Montreal Protocol. This information will be reviewed by the Conference of the Parties, at its first session and periodically thereafter, in accordance with Article 7;

(c) Calculations of emissions by sources and removals by sinks of greenhouse gases for the purposes of subparagraph (b) above should take into account the best available scientific knowledge, including of the effective capacity of sinks and the respective contributions of such gases to climate change. The Conference of the Parties shall consider and agree on methodologies for these calculations at its first session and review them regularly thereafter;

(d) The Conference of the Parties shall, at its first session, review the adequacy of subparagraphs (a) and (b) above. Such review shall be carried out in the light of the best available scientific information and assessment on climate change and its impacts, as well as relevant technical, social and economic information. Based on this review, the Conference of the Parties shall take appropriate action, which may include the adoption of amendments to the commitments in subparagraphs (a) and (b) above. The Conference of the Parties, at its first session, shall also take decisions regarding criteria for joint implementation as indicated in subparagraph (a) above. A second review of subparagraphs (a) and (b) shall take place not later than 31 December 1998, and thereafter at regular intervals determined by the Conference of the Parties, until the objective of the Convention is met;

(e) Each of these Parties shall:

  (i) coordinate as appropriate with other such Parties, relevant economic and administrative instruments developed to achieve the objective of the Convention; and
  (ii) identify and periodically review its own policies and practices which encourage activities that lead to greater levels of anthropogenic emissions of greenhouse gases not controlled by the Montreal Protocol than would otherwise occur;

(f) The Conference of the Parties shall review, not later than 31 December 1998, available information with a view to taking decisions regarding such amendments to the lists in Annexes I and II as may be appropriate, with the approval of the Party concerned;

(g) Any Party not included in Annex I may, in its instrument of ratification, acceptance, approval or accession, or at any time thereafter, notify the Depositary that it intends to be bound by subparagraphs (a) and (b) above. The Depositary shall inform the other signatories and Parties of any such notification.

3. The developed country Parties and other developed Parties included in Annex II shall provide new and additional financial resources to meet the agreed full costs incurred by developing country Parties in complying with their obligations under Article 12, paragraph 1. They shall also provide such financial resources, including for the transfer of technology, needed by the developing country Parties to meet the agreed full incremental costs of implementing measures that are covered by paragraph 1 of this Article and that are agreed between a developing country Party and the international entity or entities referred to in Article 11, in accordance with that Article. The implementation of these commitments shall take into account the need for adequacy and predictability in the flow of funds and the importance of appropriate burden sharing among the developed country Parties.

4. The developed country Parties and other developed Parties included in Annex II shall also assist the developing country Parties that are particularly vulnerable to the adverse effects of climate change in meeting costs of adaptation to those adverse effects.

5. The developed country Parties and other developed Parties included in Annex II shall take all practicable steps to promote, facilitate and finance, as appropriate, the transfer of, or access to, environmentally sound technologies and know-how to other Parties, particularly developing country Parties, to enable them to implement the provisions of the Convention. In this process, the developed country

Parties shall support the development and enhancement of endogenous capacities and technologies of developing country Parties. Other Parties and organizations in a position to do so may also assist in facilitating the transfer of such technologies.

6. In the implementation of their commitments under paragraph 2 above, a certain degree of flexibility shall be allowed by the Conference of the Parties to the Parties included in Annex I undergoing the process of transition to a market economy, in order to enhance the ability of these Parties to address climate change, including with regard to the historical level of anthropogenic emissions of greenhouse gases not controlled by the Montreal Protocol chosen as a reference.

7. The extent to which developing country Parties will effectively implement their commitments under the Convention will depend on the effective implementation by developed country Parties of their commitments under the Convention related to financial resources and transfer of technology and will take fully into account that economic and social development and poverty eradication are the first and overriding priorities of the developing country Parties.

8. In the implementation of the commitments in this Article, the Parties shall give full consideration to what actions are necessary under the Convention, including actions related to funding, insurance and the transfer of technology, to meet the specific needs and concerns of developing country Parties arising from the adverse effects of climate change and/or the impact of the implementation of response measures, especially on:

(a) Small island countries;
(b) Countries with low-lying coastal areas;
(c) Countries with arid and semi-arid areas, forested areas and areas liable to forest decay;
(d) Countries with areas prone to natural disasters;
(e) Countries with areas liable to drought and desertification;
(f) Countries with areas of high urban atmospheric pollution;
(g) Countries with areas with fragile ecosystems, including mountainous ecosystems;

(h) Countries whose economies are highly dependent on income generated from the production, processing and export, and/or on consumption of fossil fuels and associated energy-intensive products; and

(i) Landlocked and transit countries.

9. Further, the Conference of the Parties may take actions, as appropriate, with respect to this paragraph.
10. The Parties shall take full account of the specific needs and special situations of the least developed countries in their actions with regard to funding and transfer of technology.
11. The Parties shall, in accordance with Article 10, take into consideration in the implementation of the commitments of the Convention the situation of Parties, particularly developing country Parties, with economies that are vulnerable to the adverse effects of the implementation of measures to respond to climate change. This applies notably to Parties with economies that are highly dependent on income generated from the production, processing and export, and/or consumption of fossil fuels and associated energy-intensive products and/or the use of fossil fuels for which such Parties have serious difficulties in switching to alternatives.

Article 5
RESEARCH AND SYSTEMATIC OBSERVATION

In carrying out their commitments under Article 4, paragraph 1 (g), the Parties shall:

(a) Support and further develop, as appropriate, international and intergovernmental programmes and networks or organizations aimed at defining, conducting, assessing and financing research, data collection and systematic observation, taking into account the need to minimize duplication of effort;
(b) Support international and intergovernmental efforts to strengthen systematic observation and national scientific and technical research capacities and capabilities, particularly in developing countries, and to promote access to, and the exchange of, data and analyses thereof obtained from areas beyond national jurisdiction; and

(c) Take into account the particular concerns and needs of developing countries and cooperate in improving their endogenous capacities and capabilities to participate in the efforts referred to in subparagraphs (a) and (b) above.

Article 6
EDUCATION, TRAINING AND PUBLIC AWARENESS

In carrying out their commitments under Article 4, paragraph 1 (i), the Parties shall:

(a) Promote and facilitate at the national and, as appropriate, subregional and regional levels, and in accordance with national laws and regulations, and within their respective capacities:

   (i) the development and implementation of educational and public awareness programmes on climate change and its effects;
   (ii) public access to information on climate change and its effects;
   (iii) public participation in addressing climate change and its effects and developing adequate responses; and
   (iv) training of scientific, technical and managerial personnel;

(b) Cooperate in and promote, at the international level, and, where appropriate, using existing bodies:

   (i) the development and exchange of educational and public awareness material on climate change and its effects; and
   (ii) the development and implementation of education and training programmes, including the strengthening of national institutions and the exchange or secondment of personnel to train experts in this field, in particular for developing countries.

Article 7
CONFERENCE OF THE PARTIES

1. A Conference of the Parties is hereby established.
2. The Conference of the Parties, as the supreme body of this Convention, shall keep under regular review the implementation of the Convention

and any related legal instruments that the Conference of the Parties may adopt, and shall make, within its mandate, the decisions necessary to promote the effective implementation of the Convention. To this end, it shall:

 (a) Periodically examine the obligations of the Parties and the institutional arrangements under the Convention, in the light of the objective of the Convention, the experience gained in its implementation and the evolution of scientific and technological knowledge;
 (b) Promote and facilitate the exchange of information on measures adopted by the Parties to address climate change and its effects, taking into account the differing circumstances, responsibilities and capabilities of the Parties and their respective commitments under the Convention;
 (c) Facilitate, at the request of two or more Parties, the coordination of measures adopted by them to address climate change and its effects, taking into account the differing circumstances, responsibilities and capabilities of the Parties and their respective commitments under the Convention;
 (d) Promote and guide, in accordance with the objective and provisions of the Convention, the development and periodic refinement of comparable methodologies, to be agreed on by the Conference of the Parties, inter alia, for preparing inventories of greenhouse gas emissions by sources and removals by sinks, and for evaluating the effectiveness of measures to limit the emissions and enhance the removals of these gases;
 (e) Assess, on the basis of all information made available to it in accordance with the provisions of the Convention, the implementation of the Convention by the Parties, the overall effects of the measures taken pursuant to the Convention, in particular environmental, economic and social effects as well as their cumulative impacts and the extent to which progress towards the objective of the Convention is being achieved;
 (f) Consider and adopt regular reports on the implementation of the Convention and ensure their publication;

(g) Make recommendations on any matters necessary for the implementation of the Convention;
(h) Seek to mobilize financial resources in accordance with Article 4, paragraphs 3, 4 and 5, and Article 11;
(i) Establish such subsidiary bodies as are deemed necessary for the implementation of the Convention;
(j) Review reports submitted by its subsidiary bodies and provide guidance to them;
(k) Agree upon and adopt, by consensus, rules of procedure and financial rules for itself and for any subsidiary bodies;
(l) Seek and utilize, where appropriate, the services and cooperation of, and information provided by, competent international organizations and intergovernmental and non-governmental bodies; and
(m) Exercise such other functions as are required for the achievement of the objective of the Convention as well as all other functions assigned to it under the Convention.

3. The Conference of the Parties shall, at its first session, adopt its own rules of procedure as well as those of the subsidiary bodies established by the Convention, which shall include decision-making procedures for matters not already covered by decision-making procedures stipulated in the Convention. Such procedures may include specified majorities required for the adoption of particular decisions.
4. The first session of the Conference of the Parties shall be convened by the interim secretariat referred to in Article 21 and shall take place not later than one year after the date of entry into force of the Convention. Thereafter, ordinary sessions of the Conference of the Parties shall be held every year unless otherwise decided by the Conference of the Parties.
5. Extraordinary sessions of the Conference of the Parties shall be held at such other times as may be deemed necessary by the Conference, or at the written request of any Party, provided that, within six months of the request being communicated to the Parties by the secretariat, it is supported by at least one third of the Parties.
6. The United Nations, its specialized agencies and the International Atomic Energy Agency, as well as any State member thereof or observ-

ers thereto not Party to the Convention, may be represented at sessions of the Conference of the Parties as observers. Any body or agency, whether national or international, governmental or non-governmental, which is qualified in matters covered by the Convention, and which has informed the secretariat of its wish to be represented at a session of the Conference of the Parties as an observer, may be so admitted unless at least one third of the Parties present object. The admission and participation of observers shall be subject to the rules of procedure adopted by the Conference of the Parties.

Article 8
SECRETARIAT

1. A secretariat is hereby established.
2. The functions of the secretariat shall be:

    (a) To make arrangements for sessions of the Conference of the Parties and its subsidiary bodies established under the Convention and to provide them with services as required;
    (b) To compile and transmit reports submitted to it;
    (c) To facilitate assistance to the Parties, particularly developing country Parties, on request, in the compilation and communication of information required in accordance with the provisions of the Convention;
    (d) To prepare reports on its activities and present them to the Conference of the Parties;
    (e) To ensure the necessary coordination with the secretariats of other relevant international bodies;
    (f) To enter, under the overall guidance of the Conference of the Parties, into such administrative and contractual arrangements as may be required for the effective discharge of its functions; and
    (g) To perform the other secretariat functions specified in the Convention and in any of its protocols and such other functions as may be determined by the Conference of the Parties.

3. The Conference of the Parties, at its first session, shall designate a permanent secretariat and make arrangements for its functioning.

Article 9
SUBSIDIARY BODY FOR SCIENTIFIC AND TECHNOLOGICAL ADVICE

1. A subsidiary body for scientific and technological advice is hereby established to provide the Conference of the Parties and, as appropriate, its other subsidiary bodies with timely information and advice on scientific and technological matters relating to the Convention. This body shall be open to participation by all Parties and shall be multidisciplinary. It shall comprise government representatives competent in the relevant field of expertise. It shall report regularly to the Conference of the Parties on all aspects of its work.
2. Under the guidance of the Conference of the Parties, and drawing upon existing competent international bodies, this body shall:

   (a) Provide assessments of the state of scientific knowledge relating to climate change and its effects;
   (b) Prepare scientific assessments on the effects of measures taken in the implementation of the Convention;
   (c) Identify innovative, efficient and state-of-the-art technologies and know-how and advise on the ways and means of promoting development and/or transferring such technologies;
   (d) Provide advice on scientific programmes, international cooperation in research and development related to climate change, as well as on ways and means of supporting endogenous capacity-building in developing countries; and
   (e) Respond to scientific, technological and methodological questions that the Conference of the Parties and its subsidiary bodies may put to the body.

3. The functions and terms of reference of this body may be further elaborated by the Conference of the Parties.

Article 10
SUBSIDIARY BODY FOR IMPLEMENTATION

1. A subsidiary body for implementation is hereby established to assist the Conference of the Parties in the assessment and review of the effective implementation of the Convention. This body shall be open to participation by all Parties and comprise government representatives who are experts on matters related to climate change. It shall report regularly to the Conference of the Parties on all aspects of its work.
2. Under the guidance of the Conference of the Parties, this body shall:

    (a) Consider the information communicated in accordance with Article 12, paragraph 1, to assess the overall aggregated effect of the steps taken by the Parties in the light of the latest scientific assessments concerning climate change;
    (b) Consider the information communicated in accordance with Article 12, paragraph 2, in order to assist the Conference of the Parties in carrying out the reviews required by Article 4, paragraph 2 (d); and
    (c) Assist the Conference of the Parties, as appropriate, in the preparation and implementation of its decisions.

Article 11
FINANCIAL MECHANISM

1. A mechanism for the provision of financial resources on a grant or concessional basis, including for the transfer of technology, is hereby defined. It shall function under the guidance of and be accountable to the Conference of the Parties, which shall decide on its policies, programme priorities and eligibility criteria related to this Convention. Its operation shall be entrusted to one or more existing international entities.
2. The financial mechanism shall have an equitable and balanced representation of all Parties within a transparent system of governance.

3. The Conference of the Parties and the entity or entities entrusted with the operation of the financial mechanism shall agree upon arrangements to give effect to the above paragraphs, which shall include the following:

   (a) Modalities to ensure that the funded projects to address climate change are in conformity with the policies, programme priorities and eligibility criteria established by the Conference of the Parties;
   (b) Modalities by which a particular funding decision may be reconsidered in light of these policies, programme priorities and eligibility criteria;
   (c) Provision by the entity or entities of regular reports to the Conference of the Parties on its funding operations, which is consistent with the requirement for accountability set out in paragraph 1 above; and
   (d) Determination in a predictable and identifiable manner of the amount of funding necessary and available for the implementation of this Convention and the conditions under which that amount shall be periodically reviewed.

4. The Conference of the Parties shall make arrangements to implement the above-mentioned provisions at its first session, reviewing and taking into account the interim arrangements referred to in Article 21, paragraph 3, and shall decide whether these interim arrangements shall be maintained. Within four years thereafter, the Conference of the Parties shall review the financial mechanism and take appropriate measures.
5. The developed country Parties may also provide and developing country Parties avail themselves of, financial resources related to the implementation of the Convention through bilateral, regional and other multilateral channels.

Article 12
COMMUNICATION OF INFORMATION RELATED TO IMPLEMENTATION

1. In accordance with Article 4, paragraph 1, each Party shall communicate to the Conference of the Parties, through the secretariat, the following elements of information:

    (a) A national inventory of anthropogenic emissions by sources and removals by sinks of all greenhouse gases not controlled by the Montreal Protocol, to the extent its capacities permit, using comparable methodologies to be promoted and agreed upon by the Conference of the Parties;
    (b) A general description of steps taken or envisaged by the Party to implement the Convention; and
    (c) Any other information that the Party considers relevant to the achievement of the objective of the Convention and suitable for inclusion in its communication, including, if feasible, material relevant for calculations of global emission trends.

2. Each developed country Party and each other Party included in Annex I shall incorporate in its communication the following elements of information:

    (a) A detailed description of the policies and measures that it has adopted to implement its commitment under Article 4, paragraphs 2 (a) and 2 (b); and
    (b) A specific estimate of the effects that the policies and measures referred to in subparagraph (a) immediately above will have on anthropogenic emissions by its sources and removals by its sinks of greenhouse gases during the period referred to in Article 4, paragraph 2 (a).

3. In addition, each developed country Party and each other developed Party included in Annex II shall incorporate details of measures taken in accordance with Article 4, paragraphs 3, 4 and 5.

4. Developing country Parties may, on a voluntary basis, propose projects for financing, including specific technologies, materials, equipment, techniques or practices that would be needed to implement such projects, along with, if possible, an estimate of all incremental costs, of the reductions of emissions and increments of removals of greenhouse gases, as well as an estimate of the consequent benefits.
5. Each developed country Party and each other Party included in Annex I shall make its initial communication within six months of the entry into force of the Convention for that Party. Each Party not so listed shall make its initial communication within three years of the entry into force of the Convention for that Party, or of the availability of financial resources in accordance with Article 4, paragraph 3. Parties that are least developed countries may make their initial communication at their discretion. The frequency of subsequent communications by all Parties shall be determined by the Conference of the Parties, taking into account the differentiated timetable set by this paragraph.
6. Information communicated by Parties under this Article shall be transmitted by the secretariat as soon as possible to the Conference of the Parties and to any subsidiary bodies concerned. If necessary, the procedures for the communication of information may be further considered by the Conference of the Parties.
7. From its first session, the Conference of the Parties shall arrange for the provision to developing country Parties of technical and financial support, on request, in compiling and communicating information under this Article, as well as in identifying the technical and financial needs associated with proposed projects and response measures under Article 4. Such support may be provided by other Parties, by competent international organizations and by the secretariat, as appropriate.
8. Any group of Parties may, subject to guidelines adopted by the Conference of the Parties, and to prior notification to the Conference of the Parties, make a joint communication in fulfilment of their obligations under this Article, provided that such a communication includes information on the fulfilment by each of these Parties of its individual obligations under the Convention.

9. Information received by the secretariat that is designated by a Party as confidential, in accordance with criteria to be established by the Conference of the Parties, shall be aggregated by the secretariat to protect its confidentiality before being made available to any of the bodies involved in the communication and review of information.
10. Subject to paragraph 9 above, and without prejudice to the ability of any Party to make public its communication at any time, the secretariat shall make communications by Parties under this Article publicly available at the time they are submitted to the Conference of the Parties.

Article 13
RESOLUTION OF QUESTIONS REGARDING IMPLEMENTATION

The Conference of the Parties shall, at its first session, consider the establishment of a multilateral consultative process, available to Parties on their request, for the resolution of questions regarding the implementation of the Convention.

Article 14
SETTLEMENT OF DISPUTES

1. In the event of a dispute between any two or more Parties concerning the interpretation or application of the Convention, the Parties concerned shall seek a settlement of the dispute through negotiation or any other peaceful means of their own choice.
2. When ratifying, accepting, approving or acceding to the Convention, or at any time thereafter, a Party which is not a regional economic integration organization may declare in a written instrument submitted to the Depositary that, in respect of any dispute concerning the interpretation or application of the Convention, it recognizes as compulsory ipso facto and without special agreement, in relation to any Party accepting the same obligation:

    (a) Submission of the dispute to the International Court of Justice; and/or
    (b) Arbitration in accordance with procedures to be adopted by the Conference of the Parties as soon as practicable, in an annex on arbitration.

3. A Party which is a regional economic integration organization may make a declaration with like effect in relation to arbitration in accordance with the procedures referred to in subparagraph (b) above.
4. A declaration made under paragraph 2 above shall remain in force until it expires in accordance with its terms or until three months after written notice of its revocation has been deposited with the Depositary.
5. A new declaration, a notice of revocation or the expiry of a declaration shall not in any way affect proceedings pending before the International Court of Justice or the arbitral tribunal, unless the parties to the dispute otherwise agree.
6. Subject to the operation of paragraph 2 above, if after twelve months following notification by one Party to another that a dispute exists between them, the Parties concerned have not been able to settle their dispute through the means mentioned in paragraph 1 above, the dispute shall be submitted, at the request of any of the parties to the dispute, to conciliation.
7. A conciliation commission shall be created upon the request of one of the parties to the dispute. The commission shall be composed of an equal number of members appointed by each party concerned and a chairman chosen jointly by the members appointed by each party. The commission shall render a recommendatory award, which the parties shall consider in good faith.
8. Additional procedures relating to conciliation shall be adopted by the Conference of the Parties, as soon as practicable, in an annex on conciliation.
9. The provisions of this Article shall apply to any related legal instrument which the Conference of the Parties may adopt, unless the instrument provides otherwise.

Article 15
AMENDMENTS TO THE CONVENTION

1. Any Party may propose amendments to the Convention.
2. Amendments to the Convention shall be adopted at an ordinary session of the Conference of the Parties. The text of any proposed amendment to the Convention shall be communicated to the Parties by the

secretariat at least six months before the meeting at which it is proposed for adoption. The secretariat shall also communicate proposed amendments to the signatories to the Convention and, for information, to the Depositary.
3. The Parties shall make every effort to reach agreement on any proposed amendment to the Convention by consensus. If all efforts at consensus have been exhausted, and no agreement reached, the amendment shall as a last resort be adopted by a three-fourths majority vote of the Parties present and voting at the meeting. The adopted amendment shall be communicated by the secretariat to the Depositary, who shall circulate it to all Parties for their acceptance.
4. Instruments of acceptance in respect of an amendment shall be deposited with the Depositary. An amendment adopted in accordance with paragraph 3 above shall enter into force for those Parties having accepted it on the ninetieth day after the date of receipt by the Depositary of an instrument of acceptance by at least three fourths of the Parties to the Convention.
5. The amendment shall enter into force for any other Party on the ninetieth day after the date on which that Party deposits with the Depositary its instrument of acceptance of the said amendment.
6. For the purposes of this Article, "Parties present and voting" means Parties present and casting an affirmative or negative vote.

Article 16
ADOPTION AND AMENDMENT OF ANNEXES TO THE CONVENTION

1. Annexes to the Convention shall form an integral part thereof and, unless otherwise expressly provided, a reference to the Convention constitutes at the same time a reference to any annexes thereto. Without prejudice to the provisions of Article 14, paragraphs 2 (b) and 7, such annexes shall be restricted to lists, forms and any other material of a descriptive nature that is of a scientific, technical, procedural or administrative character.
2. Annexes to the Convention shall be proposed and adopted in accordance with the procedure set forth in Article 15, paragraphs 2, 3 and 4.

3. An annex that has been adopted in accordance with paragraph 2 above shall enter into force for all Parties to the Convention six months after the date of the communication by the Depositary to such Parties of the adoption of the annex, except for those Parties that have notified the Depositary, in writing, within that period of their non-acceptance of the annex. The annex shall enter into force for Parties which withdraw their notification of non-acceptance on the ninetieth day after the date on which withdrawal of such notification has been received by the Depositary.
4. The proposal, adoption and entry into force of amendments to annexes to the Convention shall be subject to the same procedure as that for the proposal, adoption and entry into force of annexes to the Convention in accordance with paragraphs 2 and 3 above.
5. If the adoption of an annex or an amendment to an annex involves an amendment to the Convention, that annex or amendment to an annex shall not enter into force until such time as the amendment to the Convention enters into force.

Article 17
PROTOCOLS

1. The Conference of the Parties may, at any ordinary session, adopt protocols to the Convention.
2. The text of any proposed protocol shall be communicated to the Parties by the secretariat at least six months before such a session.
3. The requirements for the entry into force of any protocol shall be established by that instrument.
4. Only Parties to the Convention may be Parties to a protocol.
5. Decisions under any protocol shall be taken only by the Parties to the protocol concerned.

Article 18
RIGHT TO VOTE

1. Each Party to the Convention shall have one vote, except as provided for in paragraph 2 below.

2. Regional economic integration organizations, in matters within their competence, shall exercise their right to vote with a number of votes equal to the number of their member States that are Parties to the Convention. Such an organization shall not exercise its right to vote if any of its member States exercises its right, and vice versa.

Article 19
DEPOSITARY
The Secretary-General of the United Nations shall be the Depositary of the Convention and of protocols adopted in accordance with Article 17.

Article 20
SIGNATURE
This Convention shall be open for signature by States Members of the United Nations or of any of its specialized agencies or that are Parties to the Statute of the International Court of Justice and by regional economic integration organizations at Rio de Janeiro, during the United Nations Conference on Environment and Development, and thereafter at United Nations Headquarters in New York from 20 June 1992 to 19 June 1993.

Article 21
INTERIM ARRANGEMENTS

1. The secretariat functions referred to in Article 8 will be carried out on an interim basis by the secretariat established by the General Assembly of the United Nations in its resolution 45/212 of 21 December 1990, until the completion of the first session of the Conference of the Parties.
2. The head of the interim secretariat referred to in paragraph 1 above will cooperate closely with the Intergovernmental Panel on Climate Change to ensure that the Panel can respond to the need for objective scientific and technical advice. Other relevant scientific bodies could also be consulted.
3. The Global Environment Facility of the United Nations Development Programme, the United Nations Environment Programme and the International Bank for Reconstruction and Development shall be the international entity entrusted with the operation of the financial

mechanism referred to in Article 11 on an interim basis. In this connection, the Global Environment Facility should be appropriately restructured and its membership made universal to enable it to fulfil the requirements of Article 11.

Article 22
RATIFICATION, ACCEPTANCE, APPROVAL OR ACCESSION

1. The Convention shall be subject to ratification, acceptance, approval or accession by States and by regional economic integration organizations. It shall be open for accession from the day after the date on which the Convention is closed for signature. Instruments of ratification, acceptance, approval or accession shall be deposited with the Depositary.
2. Any regional economic integration organization which becomes a Party to the Convention without any of its member States being a Party shall be bound by all the obligations under the Convention. In the case of such organizations, one or more of whose member States is a Party to the Convention, the organization and its member States shall decide on their respective responsibilities for the performance of their obligations under the Convention. In such cases, the organization and the member States shall not be entitled to exercise rights under the Convention concurrently.
3. In their instruments of ratification, acceptance, approval or accession, regional economic integration organizations shall declare the extent of their competence with respect to the matters governed by the Convention. These organizations shall also inform the Depositary, who shall in turn inform the Parties, of any substantial modification in the extent of their competence.

Article 23
ENTRY INTO FORCE

1. The Convention shall enter into force on the ninetieth day after the date of deposit of the fiftieth instrument of ratification, acceptance, approval or accession.

2. For each State or regional economic integration organization that ratifies, accepts or approves the Convention or accedes thereto after the deposit of the fiftieth instrument of ratification, acceptance, approval or accession, the Convention shall enter into force on the ninetieth day after the date of deposit by such State or regional economic integration organization of its instrument of ratification, acceptance, approval or accession.
3. For the purposes of paragraphs 1 and 2 above, any instrument deposited by a regional economic integration organization shall not be counted as additional to those deposited by States members of the organization.

Article 24
RESERVATIONS
No reservations may be made to the Convention.
Article 25
WITHDRAWAL

1. At any time after three years from the date on which the Convention has entered into force for a Party, that Party may withdraw from the Convention by giving written notification to the Depositary.
2. Any such withdrawal shall take effect upon expiry of one year from the date of receipt by the Depositary of the notification of withdrawal, or on such later date as may be specified in the notification of withdrawal.
3. Any Party that withdraws from the Convention shall be considered as also having withdrawn from any protocol to which it is a Party.

Article 26
AUTHENTIC TEXTS
The original of this Convention, of which the Arabic, Chinese, English, French, Russian and Spanish texts are equally authentic, shall be deposited with the Secretary-General of the United Nations.

IN WITNESS WHEREOF the undersigned, being duly authorized to that effect, have signed this Convention.

DONE at New York this ninth day of May one thousand nine hundred and ninety-two.

Annex I
Australia
Austria
Belarus[a]
Belgium
Bulgaria[a]
Canada
Croatia[a,*]
Czech Republic[a,*]
Denmark
European Economic Community
Estonia[a]
Finland
France
Germany
Greece
Hungary[a]
Iceland
Ireland
Italy
Japan
Latvia[a]
Liechtenstein[*]
Lithuania[a]
Luxembourg
Monaco[*]
Netherlands
New Zealand
Norway
Poland[a]
Portugal
Romania[a]
Russian Federation[a]
Slovakia[a,*]

Slovenia[a][*]
Spain
Sweden
Switzerland
Turkey
Ukraine[a]
United Kingdom of Great Britain and Northern Ireland
United States of America

[a]Countries that are undergoing the process of transition to a market economy.

[*]Publisher's note: Countries added to Annex I by an amendment that entered into force on 13 August 1998, pursuant to decision 4/CP.3 adopted at COP.3.

Annex II
Australia
Austria
Belgium
Canada
Denmark
European Economic Community
Finland
France
Germany
Greece
Iceland
Ireland
Italy
Japan
Luxembourg
Netherlands
New Zealand
Norway
Portugal
Spain
Sweden
Switzerland

United Kingdom of Great Britain and Northern Ireland
United States of America

Publisher's note: Turkey was deleted from Annex II by an amendment that entered into force 28 June 2002, pursuant to decision 26/CP.7 adopted at COP.7.

Appendix B: Climate Change Glossary

Terminology in climate change.

Climate Change: Climate change refers to a change in the state of the climate that can be identified (e.g., by using statistical tests) by changes in the mean and/or the variability of its properties, and that persists for an extended period, typically decades or longer. Climate change may be due to natural internal processes or external forcing such as modulations of the solar cycles, volcanic eruptions, and persistent anthropogenic changes in the composition of the atmosphere or in land use.

Hazard: The potential occurrence of a natural or human-induced physical event or trend or physical impact that may cause loss of life, injury, or other health impacts, as well as damage and loss to property, infrastructure, livelihoods, service provision, ecosystems, and environmental resources. In this report, the term hazard usually refers to climate-related physical events or trends or their physical impacts.

Exposure: The presence of people, livelihoods, species or ecosystems, environmental functions, services, and resources, infrastructure, or economic, social, or cultural assets in places and settings that could be adversely affected.

Vulnerability: The propensity or predisposition to be adversely affected. Vulnerability encompasses a variety of concepts and elements including sensitivity or susceptibility to harm and lack of capacity to cope and adapt.

Risk: The potential for consequences where something of value is at stake and where the outcome is uncertain, recognizing the diversity of values. Risk is often represented as probability of occurrence of hazardous events or trends multiplied by the impacts if these events or trends occur. Risk results from the interaction of vulnerability, exposure, and hazard.

Adaptation: The process of adjustment to actual or expected climate and its effects. In human systems, adaptation seeks to moderate or avoid harm or exploit beneficial opportunities. In some natural systems, human intervention may facilitate adjustment to expected climate and its effects.

Transformation: A change in the fundamental attributes of natural and human systems. Within this summary, transformation could reflect strengthened, altered, or aligned paradigms, goals, or values towards promoting adaptation for sustainable development, including poverty reduction.

Resilience: The capacity of social, economic, and environmental systems to cope with a hazardous event or trend or disturbance, responding or reorganizing in ways that maintain their essential function, identity, and structure, while also maintaining the capacity for adaptation, learning, and transformation. (Source: IPCC, 2014.)

Index

A

African Development Bank, 145
Alabama, 30
Alexandria, 185–186, 191–192
Algeria, 136
Alvarez, Isabel, 69–71
Amazon rainforest, 53
Andaman Coast, 78
Andes mountains, 65
Angola, 145
Anthropogenic, 2
Apichart, Dr. Anukularmphai, 94–95
Arab Forum for Environment and Development, 186, 191
Arctic Melting, 1, 5–6, 17, 24
Argentina, 51, 64–69
Arizona, 27, 29
Arkansas, 30
Arno river basin, 224
Assam state, 117
Atlantic City, 35

B

Baltimore, 35
Balzter, Dr. Heiko, 236–239
Bangkok, 78, 82
Bangladesh, 5, 11, 116, 164
Bangladesh Delta Plan 2100 Formulation Project, 179
Baroi, Dr. Allen Swagoto, 174–179
Bay of Bengal, 11, 84
Beijing, 110
Bihar flood, 165
Bodo, Norway, (World's first zero emission town) 2, 239–251
Boston, 27, 29, 31, 35
Botswana, 142–149
Brahmaputra river, 116–117
Brazil, 52–60
BRIC countries (Brazil, Russia, India, China), 53
Buenos Aires, 66

C

Caceras, Berta, 51
Cairo, 185–186, 189, 191
Calgary, 18
California, 27, 29, 34–35
Cambodia, 11
Canada, 15–26
Canadian Environmental Protection Act, 26
Cape Town, 135
Carter, President Jimmy, 26, 37
Centre for Landscape and Climate Research, 236, 237
Chao Phraya River, 76–77, 79
Chennai, 165–166, 170
Chief Dan George, 16
China, 3, 10–11, 22, 109–124
Chinese Academy of Sciences, 116
Climate Change Projections to 2100, 257
Climate change refugees, 23, 35, 58, 63, 67, 81–82, 86–87, 91–92, 115–116, 131–132, 140–141, 147–148, 169–170, 197, 202, 222, 227–228, 235
CO_2 emissions, 3–5
Colorado, 32
Colorado River Basin, 30
Connecticut, 31
COP21 Conference, 2
Crete, 231, 234–235
Cyclone Nargis, 84
Cyprus, 193

D

Dalai Lama, 75
Davis, Dr. Alan, 40–43
Démbelé, Dr. Moctar, 151–157
Desertification, 10, 112, 115, 128, 230–231
Durban, 135

E

East China Sea, 119
Ecuador, 60–64
Egypt, 116, 136, 185–192
Egyptian Society for Migration Studies, 190, 203
Ethiopia, 10
European Alps, 224
European Environment Agency (EEA), 225
European Union, 3, 22

F

Florida, 30, 35
Franck, Marine, 2

G

Gaborone, 143, 146
Gaborone Dam, 143, 145
Gambia, 116
Ganges river, 116
Georgia, USA, 30
Global warming, 259
GHG emissions, 2
Great Lakes, 18
Greece, 229–236
Greek Ministry of the Environment, 230
Greenland Ice Sheet (GIS), 89
Green Wall of China, 112

Guinea, 136
Guinea-Bissau, 136
Gulf of Aqaba, 199
Gyaltsen, Zamlha Tempa, 120

H

Hassan, Dr. Khaled, 203–211
Heat wave definition, 194
Houston, 16, 29
Human security, 7–8, 168
Hurricane Irma, 257
Hurricane Katrina, 16

I

Iberian Peninsula, 225–226
Illinois, 34
India, 3, 10–11, 116, 163–184
Indian Institute of Technology, 172
Indian Ocean, 126
Indonesia, 116
Indus river, 116
Inner Mongolia, 115
Intergovernmental Panel on Climate Change (IPCC), 5, 9, 51, 76, 90, 112–114, 125, 127, 141, 144, 187, 191, 200, 218, 227, 233, 260
International Food Policy Research Institute, 121
International Organization for Migration (IOM), 170
IOM, *see* International Organization for Migration
Iowa, 34
IPCC, *see* Intergovernmental Panel on Climate Change

Ireland, 220
Irrawaddy river, 116
Israel, 192–198
Italy, 223–229

J

Japan, 3, 88–93
Johannesburg, 135
Jordan, 198–203
Jordan River Basin, 193

K

Kansas, 34
Kentucky, 30
Kenya, 126–135
Kolkata, 165, 170, 258
Krabi Province, 78
Kwantlen Polytechnic University (KPU), 40
Kyoto Protocol, 10

L

La Nina, 80, 84, 129
Laos, 9
La Pampa, 66
La Plata, 66
LECZ, *see* Low elevation coastal zone
Libya, 136
Lin, Dr. J.C., 103
London, 221, 258
Los Angeles, 15–16, 29
Louisiana, 30, 32, 35
Low elevation coastal zone (LECZ), 126, 187

M

Madagascar, 11
Maharashtra flood, 165
Maine, 31
Malawi, 10, 141
Manila, 258
Marshall Islands, 5
Massachusetts, 31, 35
Mediterranean Sea, 126, 185, 189, 190, 192–193, 228, 230
Mekong river, 79, 116–117
Meridional Overturning Circulation (MOC), 7
Mexico, 15
Mexico City, 15
Miami, 29–30, 35
Migration Policy Institute, 141
Minnesota, 34
Mississippi, 30
Mombasa, 132–133, 258
Morocco, 136
Mount Aconcagua, 51
Mozambique, 135–136, 145
Mulcair, Thomas, 43–44
Mumbai, 165–66, 170–171, 258
Myanmar, 83–87
Myers, Dr. Norman, 2, 35

N

Nairobi, 127
Namibia, 141, 145
NASA, 36, 193
NASA-Goddard Institute for Space Studies, 224
National Oceanic and Atmospheric Administration (NOAA), 194
National Taiwan Normal University, 100, 103
Nebraska, 34
Nevada, 27
New Democratic Party of Canada, 43
New Hampshire, 31
New Jersey, 31, 35
New Mexico, 27, 32
New Orleans, 16, 29
New York, 15, 27, 29, 31, 35–36
New York State, 31, 35
Nigeria, 136
Nile Delta, 190
Nile river, 185–187, 189
North Carolina, 30, 34–35
North-South Water Diversion Project, China, 259
Nunavut, 23–25, 37–40

O

Ocean Acidification, 258
O'Connor, Niall, 95–100
Oklahoma, 32
Omambia, Anne Nyarichi, 149–151
Oregon, 29
Orlando, 30
Osoyoos, 19
Overland, James, 24
Oxford University Environmental Change Institute, 110

P

Palestine, 193, 202
Paris Agreement, 10, 119
Patagonia, 65
Peloponnese, 231
Pennsylvania, 31
Pericles, 257
Philadelphia, 35
Philippines, 11

Pinnerod, Ida Maria, 239–251
Plato, 1
Pope Francis, 217
Po river, 224
Port Said, 186
Puerto Rico, 27

Q

Quito, 62

R

Rahman, Dr. Md. Mizanur, 179–183
Rajagopal, Ashok Kumar, 172–174
Rasasolofoniaina, Dr. Jean Donne, 157–159
Red sea, 126, 198
Rhode Island, 31
Rio de Janeiro, 54–55, 58–59
Roosevelt, Eleanor, 260
Royal Rainmaking Project, Thailand, 78
Russia, 3

S

Salt water intrusion, 29, 223, 231
Salween river, 116–117
Santa Catarina State, 55
Scotland, 220
Sea level rise, (SLR), 6–7, 29, 88, 93, 115–116, 119, 127, 134–135, 141, 164–165, 193, 219, 224–225, 228–231
Sea of Galilee, 195
Sen, Amartya, 9
Shandong Province, 111

Shanghai, 118–119, 258
Shan State, 86
Singh, Rajendra, 163
South Africa, 134–142
South Carolina, 35
Stockholm Environment Institute (SEI), 95–97
Stockholm Water Prize, 163
Su, Dr. Shew-Jiuan, 100–103
Suez Canal, 185
Surat Thani, 77
Sutlej river, 116
Syria, 193, 202

T

Tanzania, 135
Taptuna, Peter, 37–40
Tennessee, 30
Texas, 27, 29, 32, 34–35
Thailand, 76–83
Thammasat University, 174
Thermoregulation, 234
Thessalonica, 230
Thrace, 231
Tibet, 116, 120–121
Tibet Plateau, 116
Tibet Policy Institute, 120
Tokyo, 92–93
Tokyo Bay, 92
Toronto, 15, 20
Tunisia, 136
Turkey, 193

U

Ul Hassan, Dr. Mehmood, 211–213
UK Meteorological Office, 220

UNDP, *see* United Nations Development Program
UNESCO, *see* United Nations Educational, Scientific and Cultural Organization
UNFCCC, *see* United Nations Framework Convention on Climate Change
United Kingdom, 218–223
United Nations, 191
United Nations Development Program (UNDP), 7
United Nations Educational, Scientific and Cultural Organization (UNESCO), 58
United Nations Framework Convention on Climate Change (UNFCCC), 3, 10, 26, 143, 146, 200, 219
United Nations University, 170
United States, 3, 22, 26–37
United States Department of Defense, 8
University of Leicester, 236
University of Madras, 172
Upsala Glacier, 65

V

Venice, 224, 226, 228, 258
Vermont, 31
Vietnam, 11

W

Wadhams, Peter, 24
Wajir County, Kenya, 130
Wales, 220
West Antarctica Ice Sheet, 89
West Virginia, 30
Wisconsin, 34
World Agroforestry Center, 211
World Bank, 3, 9, 79, 132, 187, 190
World Food Program (WFP), 135
World Health Organization (WHO), 9, 113, 219, 260
World Wildlife Fund, 4
Wyoming, 32

Z

Zambia, 145
Zangmu Dam, 117
Zhu, Tingju, 121–122
Zimbabwe, 10, 141

Printed by Books on Demand, Germany